# JOURNAL OF ICT STANDARDIZATION

Volume 1, No. 1 (July 2013)

# JOURNAL OF ICT STANDARDIZATION

*Chairperson:* Ramjee Prasad, CTIF, Aalborg University, Denmark
*Editor-in-Chief:* Anand R. Prasad, NEC, Japan
*Advisors:* Bilel Jamoussi, ITU, Switzerland
Jesper Jerlang, Dansk Standard, Denmark

*Editorial Board*
Kiritkumar Lathia, Independent ICT Consultant, UK
Hermann Brandt, ETSI, France
Kohei Satoh, ARIB, Japan
Sunghyun Choi, Seoul National University, South Korea
Ashutosh Dutta, AT&T, USA
Alf Zugenmaier, University of Applied Sciences Munich, Germany
Julien Laganier, Juniper Networks, USA
John Buford, Avaya, USA
Monique Morrow, Cisco, Switzerland
Vijay K. Gurbani, Alcatel Lucent, USA
Henk J. de Vries, Rotterdam School of Management, Erasmus University, The Netherlands
Yoichi Maeda, TTC Japan
Debabrata Das, IIIT-Bangalore, India
Signe Annette Bøgh, Dansk Standard, Denmark
Rajarathnam Chandramouli, Stevens Institute of Technology, USA

*Objectives:*

- Bring papers on de-jure as well as de-facto standards to the readers
- Cover pre-development, including technologies with potential of becoming a standard, as well as developed / deployed standards
- Publish on-going work with potential of becoming a standard technology
- Publish papers giving explanation of standardization process
- Publish tutorial type papers giving new comers a understanding of standardization

*Aims & Scope*

- Aim:
  - The aim of this journal is to publish standardized as well as related work making "standards" accessible to a wide public – from practitioners to new comers.
  - The journal aims at publishing in-depth as well as overview work including papers discussing standardization process and those helping new comers to understand how standards work.
- Scope:
  - Bring up-to-date information regarding standardization in the field of Information and Communication Technology (ICT) covering all protocol layers and technologies in the field

JOURNAL OF ICT STANDARDIZATION

Volume 1, No. 1 (July 2013)

| | |
|---|---:|
| Editorial Foreword | v–vi |
| N. MADELUNG and K. BERGH ANDERSEN / An Introduction to Formal Standardization and the Work on ICT Standardization in ISO/IEC – JTC1 | 1–24 |
| ROBERT M. VAN WESSEL and HENK J. DE VRIES / Business Impacts of International Standards for Information Security Management. Lessons from Case Companies | 25–40 |
| LILJANA GAVRILOVSKA and VLADIMIR ATANASOVSKI / ICT Standards in South Eastern Europe (SEE) Education: Macedonian Case | 41–58 |
| HIROSHI NAKANISHI, ROZHAN OTHMAN, KUNIO IGUSA and SHOZO KOMAKI / Global Standardization Education Program Collaborated by Osaka Univ. and MJIIT, UTM | 59–82 |
| WEIPING SUN, MUNHWAN CHOI and SUNGHYUN CHOI / IEEE 802.11ah: A Long Range 802.11 WLAN at Sub 1 GHz | 83–108 |
| JAESEUNG SONG and ANDREAS KUNZ / Towards Standardized Prevention of Unsolicited Communications and Phishing Attacks | 109–122 |

*Published, sold and distributed by:*
River Publishers
PO box 1657
Algade 42
9000 Aalborg
Denmark
Tel.: +4536953197

www.riverpublishers.com

Journal of ICT Standardization is published three times a year. Publication programme, 2013–2014: Volume 1 (3 issues)

ISSN: 2245-800X

All rights reserved © 2013 River Publishers

No part of this publication may be reproduced, stored in a retrieval system, or transmitted in any form or by any means, mechanical, photocopying, recording or otherwise, without prior written permission of the publishers.

# Editorial Foreword

Anand R. Prasad[1,2]

[1] *NEC Corporation, Japan*
[2] *Editor-in-chief Journal for ICT Standardization*

Thank you for taking interest in reading this first issue of the Journal for ICT Standardization. There is a huge gap in literature regarding material on standardization. So as to fill the gap, this Journal for ICT Standardization aims to publish standardized as well as related work making "standards" accessible to a wide public – from practitioners to newcomers and publish in-depth as well as overview work including papers discussing standardization process and those helping newcomers to understand how standards work. With the scope to bring up-to-date information regarding standardization in the field of Information and Communication Technology (ICT) covering all protocol layers and technologies in the field

Today the importance of standardization should be visible to everyone due to the penetration of mobile communications, usage WiFi or otherwise other technologies like personal computers and even electricity. Standardization makes it possible for us to communicate and devices to interoperate while reducing cost and giving choice to the customer. This all also leads to economic growth and thus overall enhancement of the society. For companies making products or providing services, standardization makes their products or services usable globally thus giving possibility to enter market worldwide. With such impact of standards or standardization on human society it is essential for people to understand standards and all aspects related to it

The objective of this journal is (1) to bring papers on de-jure as well as de-facto standards to the readers, (2) to cover pre-development, including technologies with potential of becoming a standard, as well as developed or deployed standards, (3) to publish on-going work with potential of becoming a standard technology, (4) to publish papers giving explanation of standardization process and (5) to publish tutorial type papers giving new comers a understanding of standardization

With the above objective in mind, in this first issue of the journal we bring 6 papers on standardization covering several different aspects. The first paper titled "An introduction to formal standardization and the work on ICT standardization in ISO/IEC - JTC1", by Niels Madelung and Katrine Bergh Andersen, gives an introduction to formal standardization and an overview of international standardization activities in the field of ICT. Second paper, by Robert van Wessel and Henk J. de Vries, discusses business aspects of standardization in a paper titled "Business impacts of international standards for information security management. Lessons from case companies". Followed by a third paper covering Macedonian standardization in "ICT Standards in South Eastern Europe (SEE) Education: Macedonian Case", authored by Liljana Gavrilovska and Vladimir Atanasovski, and fourth one on standardization and education in the paper titled "Global Standardization Education Program collaborated by Osaka Univ. and MJIIT, UTM" authored by Hiroshi Nakanishi, Rozhan Othman, Kunio Igusa, and Shozo Komaki. The final two papers discuss technical aspects regarding IEEE 802.11 "IEEE 802.11ah: A Long Range 802.11 WLAN at Sub 1 GHz", authored by Weiping Sun, Munhwan Choi, and Sunghyun Choi, and prevention of unsolicited communication in "Towards Standardized Prevention of Unsolicited Communications and Phishing Attacks", authored by JaeSeung Song and Andreas Kunz.

We sincerely thank the advisors of the journal and editorial board for their support. We would also like to thank the authors in trusting us with their work and River Publishers team for making this journal happen.

# An Introduction to Formal Standardization and the Work on ICT Standardization in ISO/IEC – JTC1

Niels Madelung and Katrine Bergh Andersen

*Chief consultant in Danish Standards and engaged in JTC1*
*Consultant for research in Danish Standards*

Received 22 March 2013; Accepted 14 May 2013

## Abstract

The purpose of this article is (a) to give the reader an introduction to formal standardization and (b) an overview of international standardization activities in the field of ICT.

In the first part of the article the introduction to formal standardization will be exemplified at a national level by the national Danish standardization organization (DS), at a European level by the European standardization organizations (CEN/CELENEC and ETSI) and at a global level by the international standardization organizations (ISO, IEC and ITU).

The second part of the article will provide the reader with an overview of the standards and ongoing standardization activities in JTC1 – the joint technical committee between ISO and IEC on ICT standardization. This paragraph will include a number of illustrative examples to give an in-depth impression of the work in the joint technical committee.

**Keywords:** Standardization, standards, ICT, standardization organizations, ISO/IEC JTC1, benefit from standards.

## 1 Introduction

What is a standard? Comparing different sources on the definition of the word "standard", a common understanding is "Universally or widely accepted, agreed upon, or established means of determining what something should be".

So in other words a standard needs to be established under conditions where all relevant stakeholders are involved or heard during the establishment of the standard, to insure a successful widely acceptance of the final standard.

In some cases and for a number of reasons (within ICT e.g. timeliness or limited importance or relevance), it is not feasible to aim for a widely accepted standard.

The catch 22[1] of 1) wide acceptance versus 2) swift results is in the ICT area often solved by a number of possible standards are developed rather swiftly within narrow selection of stakeholders, such as IT developing companies, and later based on (part of) such standard(s), a larger group or participants of stakeholder representatives, considers and adopt parts of such industry or commercial developed standards, into more widely accepted standards.

In other words industry or commercial developed standards are often the start of a food chain resulting into "Universally or widely accepted, agreed upon, or established means of determining what something should be".

But do we need that standards are universally or widely accepted? Not in all cases, but generally at some stage in most.

Past examples such as BETA versus VHS video tapes, resulted in a list of negative consequences for all stakeholders. The developing companies, production companies, reseller and consumers did all suffer, in terms of unnecessary time and money spend in developing, marketing and buying "the lesser desired alternative".

A more recent example is the standard for universal chargers for data-enabled mobile phones. Again millions of stakeholders have lost time and money – including the negative impact on environment.

In recent years IT developing companies have on their side involved larger and more brought representative groups of stakeholder, earlier in the industrial development of standards as well as the standardization organization e.g. ISO[2] and CEN[3], has changed the developing process of standards, to limit the timeframe for developing widely accepted standards.

---

[1] *"Catch 22" - a dilemma or difficult circumstance from which there is no escape because of mutually conflicting or dependent conditions.* (Oxford Dictionaries).

[2] The International Standardization Organization.

[3] The European Committee for Standardization.

# PART 1 – What is formal standardization? And how does it work?

In the next paragraphs we will start by explaining formal standardization and the essential characteristics of formal standardization. After that we will introduce the different standardization organizations in relation to the organizational level they represent. The standardization organizations can by organized after the area/region they cover: international standardization organizations (e.g. ISO – the International Standardization Organization, IEC – the International Electrotechnical Commission and ITU – the International Telecommunication Union), regional standardization organizations (e.g. the European standardization organizations, CEN – the European Committee for Standardization, CENELEC – the European Committee for Electrotechnical Standardization and ETSI – the European Telecommunications Standards Institute) and last but not least the national standardization organizations (e.g. DS – Danish Standards and BSI – the British Standards Institution).

Thereafter we will try to answer the question *Why are standards relevant?* by looking at the different benefits from using standards and from participating in the standardization process. Finally we have included a case from RFID (Radio Frequency Identification) standardization to show how interested parties can influence the international standardization process via their national standardization organization.

## 1.1 Formal Standardization

What is a formal standard? And what are the advantages? There are a lot of different definitions of *standard* and the term is used in a lot of different contexts. Basically you distinguish between private de facto standards and formal de jure standards. De facto standards are developed by one or more company and become standards in the extent they are used by the players in the market. Formal standards are developed in standardization organizations with formal processes for the development and approval. The formal definition of a standard is:

> *A document established by consensus and approved by a recognized body, that provides, for common and repeated use, rules, guidelines or characteristics for activities or their results, aimed at the achievement of the optimum degree of order in a given context.* (DS/EN 45020)

The formal standards have a number of essential characteristics:

**Consensus**: Standards are developed in a consensus process which means that the process is open and accessible for all parties interested, and all interest are attempted met through agreement. Standards are in other words not necessarily an expression of the highest expertise or the highest requirements but rather an expression of what the parties involved could agree upon.

**Voluntariness**: The standardization organizations publish the standards, but they are voluntary to use and the standardization organizations are not obligated to control whether the standards are used correctly. It is therefore crucial for the use of the standards that they are considered relevant by the users – a relevance that often is established in this consensus process.

**Contractual basic:** Standards are widely used as part of the contractual basic in for example tenders or other contracts. This means that standards frequently become market requirements and from that point on are not experienced as voluntary. This is also the case when standards are mentioned in the law e.g. in directives from the European Commission.

There are a number of different types of standards each with different purposes:

- Terminology standards, which determine terms in different areas in order to make people "speak the same language"
- Product standards, which describe specifications, requirement to functionality or safety for a product
- Test standards, which determine how you test a product in accordance with requirements to functionality
- Standards for measurement, which specify how you measure e.g. concentration of toxic substances in water or in the air.
- Management standards, which are used to control organizations or companies in regard to quality, environmental management, energy performance etc. They are typically used as a basic for certification.

It is not always possible or necessary to make a formal standard. Therefore the standardization organizations have developed other types of documents – a kind of preliminary standard. This is for example "technical specifications" and "international workshop agreements". Another example is a "guideline" which is sometimes made to support a standard – e.g. if the content is very difficult or if there are several ways to interpret a standard a guideline might be handy.

Sometimes the standardization process takes too long for a fast moving industry or product development. These other formal documents can also be a "fast track" to a standard by shortening the standardization process and making the requirements/test method/etc. faster available for the users.

### 1.1.1 The Standardization Process

The standardization process is characterized by formal processes for the development and approval of standards. In short it is summarized in figure 1 below for an ISO standard. The process is very similar in other formal standardization organizations with few exceptions.

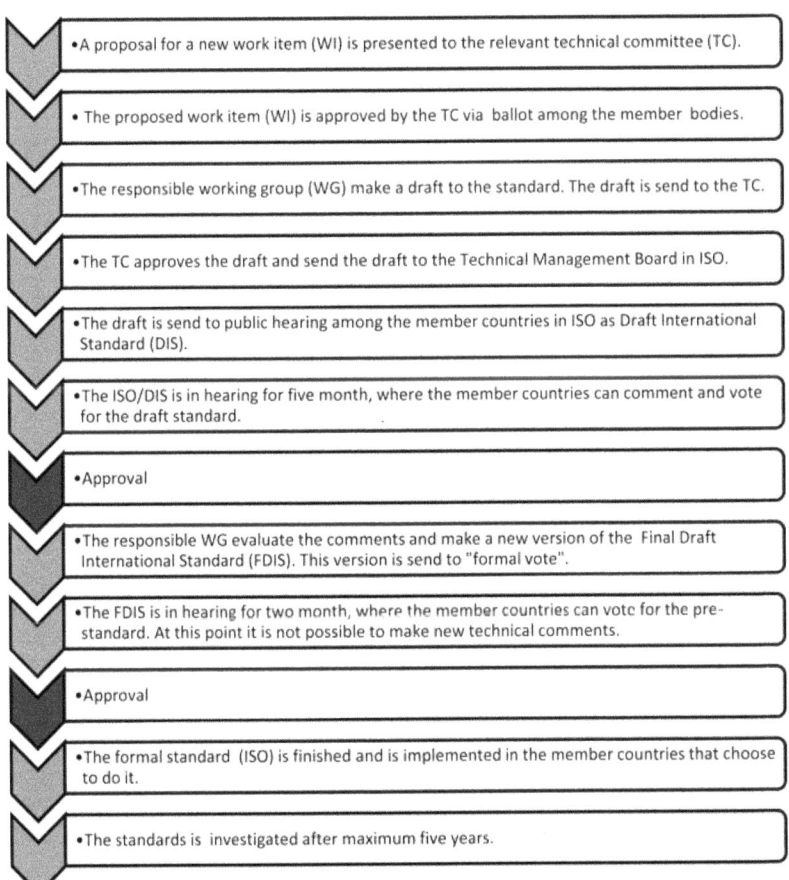

Figure 1 The development of an ISO standard.

## 1.2 International Standardization: ISO, IEC and ITU

In popular terms standardization is said to be the largest network in the world with around 100.000 members in ISO[4] even more if we include the members in IEC[5] and ITU[6]. Standardization is a continuous everyday activity – actually an average of eight technical meetings in ISO takes place every working day.

ISO and IEC are similar in a lot of ways. IEC covers the electrotechnical standardization and ISO roughly the rest. ISO and IEC are member organizations with one member organization from each member country. Danish Standards is the national Danish member of ISO and IEC – Danish companies, organizations, research institutions, government etc. that wants to participate in the work in ISO or IEC needs to be member of Danish Standards. 163 countries are member of ISO and 82 of IEC – all these countries are represented by a national standardization body – e.g. British Standards Institute in United Kingdom.

ITU differs from ISO and IEC in some ways. It is United Nations specialized agency for information and communication technologies. Being a UN-agency ITU is managed by a different set of rules than ISO and IEC. For example it is possible for a company, organization or research institution to be member of ITU directly – without being member of the national standardization body.

You can read more about the specific organizations on their web pages:

- www.iso.org
- www.iec.ch
- www.itu.org

## 1.3 European Standardization: CEN/CENELEC and ETSI

Besides the international level standardization can also be organized within regions e.g. Europe or the Northern countries/Scandinavia. In this section we will focus on the European standardization organizations; CEN (the European Committee for Standardization), CENELEC (the European Committee for Electrotechnical Standardization) and ETSI (the European Telecommunications Standards Institute).

CEN corresponds to ISO at international level, where CENELEC corresponds to IEC and ETSI to ITU.

---

[4] International Standardization Organization.
[5] International Electrotechnical Commission.
[6] International Telecommunication Union.

**CEN** is responsible for all formal standardization in Europe except for the electrotechnical (CENELEC) and telecommunication (ETSI). CEN was funded back in 1961 as the first European standardization organization. Today CEN has 33 member countries which make the number of participating countries bigger than in the European Union.

For further information see www.cen.eu.

**CENELEC** is responsible for the formal electrotechnical standardization in Europe. As CEN, CENELEC has 33 member countries within Europe.

For further information see www.cenelec.eu.

Together CEN and CENELEC have around 13.000 published standards (EN-standards) even more if you count related documents like technical specifications etc.

**ETSI** was established in 1988 by CEPT (European Conference of Postal and Telecommunication Administrations) on the basis of a proposal from the European Commission that recommended the funding of an organization with responsibility for the establishing of common European standards telecommunication. Today ETSI is responsible for the standardization of information and communications technologies.

## Influence/membership of the European standardization organizations

Danish Standards is the national member of CEN, CENELEC and ETSI for Denmark. Danish Standards is therefore responsible for representing the Danish interests in the three organizations and for providing Danish companies, NGO's, authorities and other interested parties access to the standardization work in the three organizations.

There is a little but very essential difference between CEN/CENELEC and ETSI regarding membership. In CEN and CENELEC a company or another interested party can only be member/participate in the standardization process through membership of a national standardization organization – e.g. Danish Standards. In ETSI it is also possible for a company or another interested party to participate through a national standardization organization, but in ETSI it is also possible for a company or another interested party to be member directly by paying an annual fee. For further information see www.etsi.org.

### 1.3.1 The Link Between the European Organizations

The collaboration between the European standardization organizations are very strong, they have a lot of joint working groups on different transverse subjects e.g. education about standardization and engaging SMEs. The collaboration between CEN and CENELEC are particularly strong and they decided in 2010 to establish a common management centre – CEN-CENELEC Management Centre (CCMC) in Brussels.

See www.cencenelec.eu for further information.

All member countries in CEN/CENELEC are obliged to implement all published standards (EN-standards) as national standards and to withdraw any conflicting national standards (if any). This is defined in the internal regulation in CEN/CENELEC. This principle strengthen the European standardization by establishing a common ground and is therefore of high importance.

## 1.4 National Standardization: Danish Standards

Last but not least we have the national level in standardization – it is here that all the companies, experts, researchers etc. are member and get their influence (except with ITU and ETSI see previous paragraphs).

Almost every country in the world has a formal national standardization body – some countries even have two because they have a separate body to handle the electrotechnical standardization. In figure 2 below your find an illustration of the extension of standardization worldwide exemplified in a map of ISO[7] member countries.

Figure 2 All the gray countries have a formal standardization body[7]. (The different tones of gray indicate the status of their membership in ISO.)

---

[7] Greenland does not have a formal national standardization body – in practice they have a close cooperation with Danish Standards to ensure the use of standards in Greenland and Greenland's interests.

Danish Standards is the formal standardization body in Denmark. We are therefore responsible for the formal standardization in Denmark and for representing the Danish interest in European and international standardization – we are therefore the national member of ISO, IEC, CEN, CENELEC and ETSI. Danish Standards is not member of ITU – here Denmark is represented by the national authorities.

Danish Standards has around 220 technical committees spread over the areas where we have interested stakeholders in Denmark. There are almost 2000 members of our technical committees from around 850 different companies.

Danish Standards is also responsible for the Danish WTO[8] Enquiry Point. We answer 4000 to 5000 enquiries every year from Danish and foreign companies and organizations about standards and trade with Denmark.

## 1.5 Why Standards?

A very important and relevant question to ask at this point is WHY standards/standardization? What are the benefits? In the following sections we will specify the benefits from using standards or participating in the standardization process for different segments.

The general philosophy behind standardization is to make life easier for the users of the standards by making it easier to specify e.g. requirements to a product (they are specified in the standard) and by making it easier for the company to document that they fulfil the requirements. This is done by establishing a common ground in the standards for e.g. test requirements or quality management. Depending on the standard a company using a standard makes sure that their product or service is compatible with other products and meets the requirements in the law (e.g. the use of harmonised standards regarding CE-marking in the European Union).

Standards also enhance market maturity and can be used to enter new markets to develop growth. This is relevant because a standard can help maturing a technology to be ready for the market. This is mainly the case with new standardization topics.

When participating in standardization you basically get three things: knowledge, influence and network.

**Knowledge** of the upcoming standards and the underlying rationales.

**Influence** on the upcoming standards and the requirements in them.

---

[8] World Trade Organization.

**Network** with the other participants, often from different positions in the society than yourself and from around the world. In popular terms standardization is said to be the biggest network in the world. This network will also contribute to your company or organization by providing knowledge and influence to e.g. the product development.

### 1.5.1 A Society Point of View

Looking at the benefits from a society point of view the following benefits can be mentioned. Standards:

- are a catalyst for growth
- increase international trade (by reducing trade barriers and making the transition of goods between companies/markets easier)
- utilize the expert knowledge in society/industry and best practices – to make the knowledge accessible for companies and society in general
- enhance the innovation in the companies – by setting new standards they need to improve/develop
- define a base for developing and implementing new technologies/knowledge – e.g. smart-grid and electrical cars.

### 1.5.2 A Company Point of View

Besides the previous mentioned benefits which of course also apply for companies, standards are a highly efficient means to inject market knowledge directly into the company's product development, service design, and other innovative processes. It is also relevant the other way around – companies can use standards to disseminate knowledge to the market in order to make their way "the way".

Standards can also be used to give directions/requirements to a subcontractor in order to improve the probability for receiving the desired product – maybe to ensure compatibility with existing products or components. It is much easier for a company to refer to a standard with requirements than to formulate the requirements from time to time. Making their own requirements also enhances the probability for misunderstandings. When using a formal standard it is more likely to avoid misunderstandings and misinterpretations because of the dissemination and accept of formal standards. For start-up companies, and for any company that wants to launch a new product in a given market, it is a major advantage that they know how to use and benefit from the system of standardization.

Standards are often key to market access – they help ensure compatibility, safety, and compliance with regulatory and market demands. In other words standards minimize the time-to-market for new inventions because they can be used to pave the way by removing barriers. Standards are also known for minimizing the costs in intra-Community trade.

Standards can literally make or break the successful commercialization of a new idea. An example of this could be a company that has developed a new product for the market (e.g. a piece of electrical equipment). First when the product is fully developed and ready to go to the market the company finds out that the product needs to be CE-marked. To CE-mark electrical equipment the product needs to fulfil the requirements in the relevant European directive (2006/95/EC) and the relevant harmonised standards. If the product does not fulfil the requirements in the harmonised standards and has not been tested after the standardized test methods, the product is considered to be illegal and the product development process needs to start over – now taking the relevant standards into account. Not knowing the right standards can, especially for a small company, mean the difference between bankruptcy and success.

An analysis from DAMVAD (a northern consultancy bureau) made for The Danish Business Authority published in 2013 have looked at the value creation of standards in the companies using them. Companies most often use/implement standards because of a demand from the market or a requirement in the law, but the analysis shows that they also experience unexpected or other benefits:

- 77 % of the companies in the analysis experience that standards enhance the quality of their product
- 80 % of the companies experience that standards contribute with new knowledge
- 60 % experience that the co-operation with subcontractors and customers become more simple
- 72 % experience that implementing standards increase the confidence from the customers.

### 1.5.3 A Research Point of View

The last perspective we will look at regarding benefits from using standards and participating in standardization is from a research point of view. The focus on using standards in connection with dissemination of research results are increasing, especially in Europe where the focus on dissemination research results to get funding from the European framework programmes (FP7 and

Horizon2020) are getting stronger. According to the newest draft version of Horizon2020 standardization will be mentioned in Horizon2020[9] as an aspect to consider in the dissemination phase. Therefore applications including standardization will be more likely to get funding.

The focus on using standards as dissemination of research results come from the importance for the society, the industry, growth and employment that the result from research projects are communicated and made available for the public. Standardization is one way of ensuring this – and is getting increasingly more widespread. Standard *open up* the knowledge and is making the knowledge available to all interested parties in contrast to e.g. patenting and academic articles (often for a limited audience).

### 1.5.4 A Political Push for ICT Standardization

Standardization has different impact in different areas – some of it is linked to the spreading and number of standards with in the area. The above-mentioned benefits from standards and standardization more or less also apply for ICT-standardization. Very relevant for the ICT area is the maturing of markets and technologies at the same time. This convergence lead to a more constant development and the companies therefore get a longer period for investment – this again increases the investments and creates a basis for research and development activities, which increases the competitiveness.

To zoom on the European level there are an increasingly focus on the necessity for standardization within ICT. This is for example something The European Commission's recognizes the importance of in their 2020-strategy:

> "At EU level, the Commission will work:
> 
> - …
> - *To reform the research and innovation funds and increase support in the field of ICTs so as to reinforce Europe's technology strength in key strategic fields and create the conditions for high growth SMEs to lead emerging markets and to stimulate ICT innovation across all business sectors;"*
> 
> (*Europe 2020 – A European strategy for smart, sustainable and inclusive growth*, EC, Communication from the Commission, 2010)

---

[9] When writing this article Horizon2020 was not finally approved but still for negotiation in the European Commission.

*An Introduction to Formal Standardization* 13

This is also reflected in a new regulation from the European Parliament on European Standardization, where it is made clear, that ICT standardization is a priority (Chapter IV).

Therefore standardization with ICT is an important focus area in the relevant standardization organizations these years.

## 1.6 From a National to an International Standard – a Case of the RFID Chip

It might seem to be very difficult to influence the development of international standards - but that is not the case. As this case will illustrate you just need to be strategic, willing to negotiate and aware of the political game in standardization.

The following case is based on an interview with Carsten Riis Fredriksen, standardization consultant in Danish Standards and working with RFID[10] standardization.

### 1.6.1 Development if "The Danish Data Model"

In the beginning of the twenty-first centuries The Danish Agency for Culture decided that they wanted a standard for RFID microchips for the libraries. In cooperation with Danish Standards a national committee was established. From the beginning a lot of different partners were involved, Danish as well as international, because The Danish Agency for Culture wanted an internationally valid standard. The reason that they wanted an internationally valid standard was to create a platform for the suppliers to deliver the RFID system for the Danish libraries. This would help create a market for RFID chips for libraries and to create a system applicable for all libraries in Denmark. Other countries would benefit from this as well.

In 2005 they succeeded in making a (national Danish) technical information called DS/INF 163, RFID Data Model for Libraries.

### 1.6.2 From a National to an International Standard

From this point on the goal was to obtain international agreement on the standard in order to make an ISO standard. A working group under ISO TC 46/SC 4 Technical interoperability was established and a representative from Denmark became convener for the working group that was to make the standards for RFID in libraries (ISO/TC 46/SC 4/WG 11).

---

[10] Radio Frequency Identification.

Some countries were against "The Danish Data Model" because it did not fit their national systems. To meet this demands the Danish team came up with the idea of making one overall standard with the general guidelines and to separate parts – one in accordance with The Danish Data Model and one in accordance with the other system. The model with one overall standard and to part for the two different systems made it possible to reach consensus. The three ISO standards compose the ISO 28560-series that were published in 2011.

### 1.6.3 The National Situation Today

When the ISO 28560-series on RFID was approved the national technical information in Denmark – DS/INF 163 – was withdrawn because it had become unnecessary. To make the implementation of the ISO 28560-series easy for the Danish libraries a new technical information was made specifying the specific Danish requirements (DS/INF 28560).

RFID not only used in libraries but in a wide variety of businesses where there is a need for information exchange e.g. traceability of containers or medical equipment. Denmark is also active in other areas of RFID, but it is only in the case of libraries that our influence has been that extensive.

---
**Facts**
RFID - Radio Frequently Identification - is a small microchip that, when activated, can send information (e.g. the identification code for a book) to a receiver.
The libraries get a lot of advantages by using RFID chips in their books. They can for example easily find dislocated books or register several books at the same time.

---

### 1.6.4 The Standards

DS/INF 163        RFID Data Model for Libraries (withdrawn)
DS/ISO 28560-1    Information and documentation – RFID in libraries
                  – Part 1: Date elements and general guidelines
                  for implementation
DS/ISO 28560-2    Information and documentation – RFID in libraries
                  – Part 2: Encoding of RFID date elements based on
                  rules from ISO/IEC 15962
DS/ISO 28560-3    Information and documentation
                  – RFID in libraries – Part 3: Fixed length encoding
DS/INF 28560      RFID in libraries

# PART 2 – A Snapshot of the International Standardization Activities in ICT

## 2.1 The Background

ISO started its ICT standardization activities in the beginning of 1960, back in the days when a computer should have the size of a small office building to match 5 % of the power of a laptop anno 2013.

ICT took off both from a technical and a user point-of-view. Both ISO and IEC worked hard on keeping up with the increasing demand of standards. By the 1980'ies computer were everywhere from business, government, schools and private homes.

To keep up the constant increasing demand of standards and to avoid the risk of duplicative or incompatibly standards, ISO and IEC joint forces in the area of ICT in 1987, by joining their respectively technical committees (all together three) into what today is known as the joint technical committee ISO/IEC JTC 1.

In 2012, JTC 1 celebrated 25 years of ICT International Standards development (1987 - 2012). Read "25 years of JTC 1 - We've come a long way!" for more information.

Today, ISO/IEC JTC 1 is one of the largest and most prolific technical committees in international standardization. With over 2.500 published standards under the broad umbrella of the committee and its 19 subcommittees, ISO/IEC JTC 1 makes a huge impact on the ICT industry worldwide.

By 2013 JTC 1 consists of more than 100 working groups, under its 19 subcommittees. It's within the working groups that standards are developed.

The reason for that we in our day-to-day activities can make "sense" out of the characters on our keyboards, create MPEG files, use smartcards and much more, is thanks to the dedicated work of the experts involved in JTC 1 working groups, representing close to half of the countries in the world.

### 2.1.1 The Mission and Principles

#### Mission

JTC 1s mission is to develop, maintain, promote and facilitate IT standards required by global markets meeting business and user requirements concerning:

- Design and development of IT systems and tools
- Performance and quality of IT products and systems
- Security of IT systems and information

- Portability of application programs
- Interoperability of IT products and systems
- Unified tools and environments
- Harmonized IT vocabulary
- User friendly and ergonomically designed user interfaces

**Principles**

JTC 1 standards development will be conducted with full attention to a strong business-like approach (e.g., cost effective, short development times, market-oriented results.)

JTC 1 will provide a wide range of quality products and services, within its scope and mission, to cover identified global needs.

The JTC 1 community will actively promote the use of its products and services and the timely implementation of JTC 1 standards within the form of useful products on a worldwide basis.

JTC 1 will ensure that its user needs including multicultural requirements, are fully met, such that its products and services promote international trade.

JTC 1 recognizes the value of the work of other organizations and the contribution they make to international IT standardization and will complement existing and forthcoming JTC 1 programs through other leading edge activity with the objective of providing the best standards worldwide.

JTC 1 will provide a standards development environment which attracts technical experts and users having identified standardization needs.

## 2.2 The Structure

ISO/IEC JTC 1 acts as an umbrella and management board for a number of SC (sub-committees).

JTC 1 has a chairperson (currently Ms. Karen Higgenbottom, USA) and conduct annual meetings – JTC 1 Plenary meetings - where countries participating in the work of JTC 1's SC participates, receives reports and request from the chairpersons of the individual SC and make decisions in terms of initiating and delegating of work, cancellation of work, establishment and closing of sub-committees etc.

All SC's and WG's conducts there work in cooperation with other standardization organizations (e.g. technology or industry oriented) where relevant, to avoid conflicts or duplication of work.

Much cooperation is long lasting and the exchange of information is formalized by entering into a liaison agreement and by each party appointing

*An Introduction to Formal Standardization* 17

liaison officers that site-in on the other parties meetings, as well as being included on mailing list etc.

Especially during the last 10 years development and practical use of ICT, the "old" definitions and silos in which we use to divide ICT into has been challenged. E.g. Smart Grid is a good example, since both the power technology, power meters, information security and other ICT aspects goes into that new and growing technology and business area. Other examples on the falling down of silos are Augmented Reality and Cloud Computing, the latter also motivated by the privatisation of the National Telco Companies.

For those reasons the following names (definitions) must be taken with a grain of salt. And for that reason as well the number of liaison partnerships is steadily growing and branching out.

Table 1 JTC 1 working groups and sub-committees with underlying working groups.

| SC/WG | Name and Working Groups |
|---|---|
| SWG 1[11] | **Accessibility** |
| AHG 3[12] | **Tools** |
| AHG 2 | **Structure** |
| SWG 3 | **Planning** |
| SWG 2 | **SWG – Directives** |
| AHG 1 | **Incubator** |
| WG 7 | **Sensor networks** |
| WG | **Governance of IT** |
| SC 2 | **Coded character sets** |
| | WG 2 Universal coded character sets |
| SC 6 | **Telecommunication and information exchange between systems** |
| | WG 1 Physical and data link layers |
| | WG 7 Network, transport and future network |
| | WG 8 Directory |
| | WG 9 ASN.1 and registration |
| SC 7 | **Software and systems engineering** |
| | SWG 5 Standards management group |
| | SWG 22 Vocabulary validation |
| | AG 1[13] Life Cycle Processes Harmonization |
| | WG 2 System software documentation |
| | WG 4 Tools and environment |
| | WG 6 Evaluation and metrics |
| | WG 7 Life cycle management |
| | WG 10 Process assessment |

---

[11] SWG – Special Working Group.
[12] AHG – Ad-Hoc Group.
[13] AG – Advisory Group.

Table 1 Continued

| SC/WG | Name and Working Groups |
|---|---|
| | WG 19 Open distributed processing and modelling languages |
| | WG 20 Software and systems bodies of knowledge and professionalization |
| | WG 21 Software asset management |
| | WG 23 System quality management |
| | WG 24 SLC Profile and guidelines for VSE |
| | WG 25 IT Service management |
| | WG 26 Software testing |
| | WG 27 IT enabled services |
| | WG 28 Common Industry Formats for Usability Reports |
| | WG 42 Architecture |
| SC 22 | **Programming languages, their environments and system software interfaces** |
| | WG 4 COCOL |
| | WG 5 Fortran |
| | WG 9 Ada |
| | WG 14 C |
| | WG 17 Prolog |
| | WG 21 C++ |
| | WG 23 Programming Language Vulnerabilities |
| SC 23 | **Digital Recorded Media for information interchange and storage** |
| | WG 6 iVDR Cartridge |
| | WG 7 Joint WG between SC 23, TC 42 and TC171/SC 1 |
| SC 24 | **Computer graphics, image processing and environmental data representation** |
| | WG 6 Augmented reality continuum presentation and interchange |
| | WG 7 Image processing and interchange |
| | WG 8 Environmental representation |
| | WG 9 Augmented reality continuum concept and reference model |
| SC 25 | **Interconnection of information technology equipment** |
| | WG 1 Home electronic systems |
| | WG 3 Customer premises cabling |
| | WG 4 Interconnection of computer systems and attached equipment |
| SC 27 | **IT Security techniques** |
| | WG 1 Information security management systems |
| | WG 2 Cryptography and security mechanisms |
| | WG 3 Security evaluation, testing and specification |
| | WG 4 Security controls and services |
| | WG 5 Identity management and privacy technologies |
| SC 28 | **Office equipment** |
| | WG 1 Advisory WG |
| | WG 2 Consumables |
| | WG 3 Productivity |
| | WG 4 Image quality assessment |
| | WG 5 Office Colour |

Table 1 Continued

| SC/WG | Name and Working Groups |
|---|---|
| SC 29 | **Coding of audio, picture, multimedia and hypermedia information** |
| | WG 1 Coding of still pictures |
| | WG 11 Coding of moving pictures and audio |
| SC 31 | **Automatic identification and data capture techniques** |
| | WG 1 Data carrier |
| | WG 2 Data structure |
| | WG 4 Radio frequency identification for item management |
| | WG 5 Real time locating systems |
| | WG 6 Mobile item identification and management |
| | WG 7 Security for item management |
| SC 32 | **Data management and interchange** |
| | WG 1 eBusiness |
| | WG 2 MetaData |
| | WG 3 Database language |
| | WG 4 SQL/Multimedia and application packages |
| SC 34 | **Document description and processing languages** |
| | WG 1 Information description |
| | WG 2 Information presentation |
| | WG 3 Information association |
| | WG 4 Office Open XML |
| | WG 5 Document interoperability |
| | WG 6 OpenDocument Format |
| SC 35 | **User interface** |
| | WG 1 Keyboards and input interfaces |
| | WG 2 Graphical user interface and interaction |
| | WG 4 User interfaces for mobile devices |
| | WG 5 Cultural and linguistic adaptability |
| | WG 6 User interfaces accessibility |
| | WG 7 User interfaces object, actions and attributes |
| | WG 8 User interfaces for remote interactions |
| SC 36 | **Information technology for learning, education and training** |
| | WG 1 Vocabulary |
| | WG 2 Collaborative technology |
| | WG 3 Learner information |
| | WG 4 Management and delivery of learning, education and training |
| | WG 5 Quality assurance and descriptive frameworks |
| | WG 6 Platform, Services and Specification Integration |
| | WG 7 ITLET – Culture, language and individual needs |
| SC 37 | **Biometrics** |
| | WG 1 Harmonized biometric vocabulary |
| | WG 2 Biometric technical interfaces |
| | WG 3 Biometric data interchange formats |

Table 1 Continued

| SC/WG | Name and Working Groups |
|---|---|
|  | WG 4 Biometric functional architecture and related profiles |
|  | WG 5 Biometric testing and reporting |
|  | WG 6 Cross-Jurisdictional and Social Aspects of Biometrics |
| SC 38 | **Distributed application platform and services** |
|  | WG 1 Web services |
|  | WG 2 Service Oriented Architecture (SOA) |
|  | WG 3 Cloud Computing |
| SC 39 | **Sustainability for and by information technology** |
|  | WG 1 Resource Efficient Data Centres |
|  | WG 2 Green ICT |

## 2.3 The Outcome

The number of projects on new standards and revision of current ones is very high. In some SC they have a "wish list" of projects that they have to hold back due to the load of current projects. This is also a subject that often is brought up on the JTC 1 Plenaries, where the load of project are tried to be spread out.

The following acts only as examples on what lately has been or is in the process of being produced by the SC under JTC 1.

Table 2 Current projects under JTC 1

| SC | ISO/IEC standard number and title (part of) |
|---|---|
| 2 | 10646 Revised Universal Character Set |
|  | 10646 Character for Old Hungarian, Albanian and others |
| 6 | 9834 Generation of Universally Unique Identifiers (UUID) |
|  | 15149 Magnetic field area network (MFAN) |
| 7 | 30105 Business Process Outsourcing Lifecycle Processes |
|  | 42030 Architecture Evaluation |
|  | 90003 Application of ISO 9001 to computer software |
| 17 | 11694 Use of biometrics on an optical memory card |
|  | 18745 Physical Test Methods for Passport Books |
|  | 30117 Guide to on-card biometric comparison standards and appl. |
| 23 | 29121 Data migration method for DVD-R, -RW, -RAM and others |
| 24 | 19775 Extensible 3D (X3D) – Processing (several standards) |
|  | 19776 Extensible 3D (X3D) – Representation (several standards) |
| 25 | 29104 Management protocol for ubiquitous home network services |
|  | 29108 Terminology for intelligent homes |
|  | 29145 Wireless Beacon-enabled Energy Efficient Mesh network |
|  | 30100 Home network management |

Table 2 Continued

| SC | ISO/IEC standard number and title (part of) |
|---|---|
| 27 | 18370 Blind digital signatures |
|  | 20008 Anonymous digital signatures |
|  | 24760 Framework for Identity Management |
|  | 27001 Information security management systems[14] |
|  | 27016 Organizational economics |
|  | 27017 Information security controls for the use of Cloud Computing |
|  | 27018 Data protection controls for public cloud computing services |
|  | 27035 Information security incident management |
|  | 27036 Information security for supplier relationships |
|  | 29101 Privacy architecture framework |
| 29 | 15444 JPEG 2000 image coding systems Compound image file form. |
|  | 23000 MM appl. Format (MPEG-A) Augmented reality appl. form. |
| 31 | 15961-2 RFID for item management Registration of RFID data |
|  | 15961-3 RFID for item management RFID data constructs |
|  | 18004 Automatic Id and data capt. tech. QR bar code symb. spec. |
| 32 | 19763 MFI Core model and basic mapping |
| 34 | 30114 Extensions of office open XML file format Guidelines |
| 35 | 30109 Worldwide-available personalized computing environment |
|  | 30113 Gesture-based interface across device and methods |
|  | 30122 Principal voice commands Framework and general guidance |
| 36 | 19788 Learning, education and training Metadata for learning res. |
|  | 19796 Quality for learning, educ., and training Products and Serv. |
|  | 36001 Quality for learning, educ., and training Management Syst. |
| 37 | 19785 Common Biometric Exchange Formats Framework |
|  | 19794 XML Encoding |
|  | 29196 Guidance for Biometric Enrolment |
|  | 29794 Biometric Sample Quality Standard Iris image |
|  | 30107 Anti-Spoofing and Liveness Detection Techniques |
| 38 | 17788 Cloud Computing Reference Architecture |
| 39 | 30132 IT sustainability Guidance for dev. of Energy Efficient ICT |
|  | 30134 Data Centres KPI Power usage effectiveness |

## 3 Conclusion and Perspectives

The purpose with this article was to introduce formal standardization – its benefits and advantages – plus to give an overview of activities within ICT standardization in ISO/IEC JTC1. The intention was to strengthen the necessity and the benefits from knowing and using formal standardization. The focus on formal standardization is in many contexts increasing and the benefits for companies, society and researchers are getting more explicit. As the devel-

---

[14] As from 2013 all parts of the Danish Government has to comply with this standard.

opment is at the moment, standardization will only become more widespread over the next years. We hope this article will provide the reader with a taste of the formal standardization world both in general and within ICT.

The importance of what we today refer to as ICT standardization will go on another 25+ years. The job has not been done yet and never will. The scene of ICT grows bigger and more interrelated with other part of what we used to see at completely separate issues.

Who ever thought 10 years ago that our electrical power should be connected with ICT, or our legal right of privacy had anything to do with ICT, (mobile) phones or driving and navigating?

The many coming years challenge concerning ICT, will be the interrelated and innovative new ways of conducting our business and private lives local and international.

The good old days where we neatly could separate things under each completely separate label has long gone, and we are now - without knowing it – floating around, look for new labels.

Maybe 5-10 years from now, we will not refer to ICT as common nominator, but will have invented completely new ways of separating logical groups from each other.

But whatever new areas we invent or new labels we will use, standardization will not be less important. On the contrary "Universally or widely accepted, agreed upon, or established means of determining what something should be" will become even more important.

The timeframe of creating standards will play an even important role as well will the agility in revising standards. The gallery of stakeholders will develop; the users will get more directly involved and have a larger impact of what will be "accepted" and what not.

ICT standardization is here to stay, it whatever name, shape or form – that is already universally agreed.

## References

[1] *25 years of ISO/IEC JTC 1 - We've come a long way!* http://www.iso.org/iso/home/news_index/news_archive/news.htm?refid=Ref1601
[2] DS/EN 45020, *Standardization and related activities - General vocabulary*
[3] DS-hæfte 17, *Introduktion til standardisering*, Danish Standards, 2012
[4] *Europe 2020 – A European strategy for smart, sustainable and inclusive growth*, EC, Communication from the Commission, 2010 http://ec.europa.eu/eu2020/pdf/COMPLET%20EN%20BARROSO%20%20%20007%20-%20Europe%202020%20-%20EN%20version.pdf

[5] Jerlang, Jesper, "Standardisering som regulering", *Standardisering som regulering – en antologi om standardisering i krydsfeltet mellem politik og marked*. DS-hæfte 32, Danish Standards 2012, pp. 13-27.
[6] REGULATION (EU) No 1025/2012 OF THE EUROPEAN PARLIAMENT AND OF THE COUNCIL of 25 October 2012 on European standardisation (Chapter IV) http://eur-lex.europa.eu/LexUriServ/LexUriServ.do?uri=OJ:L:2012:316:0012:0033:EN:PDF
[7] *Standarder som værdiskaber i danske virksomheder*. Report made by DAMVAD for The Danish Business Authority (January 2013)

## Biography

Katrine Bergh Andersen
Consultant for research and education in Danish Standards
Katrine Bergh Andersen is working with strengthening the relation between standardization and research, which she has been doing since 2010. Before that she has been working as information consultant in Danish Standards. She has a master's degree in psychology of language from the University of Copenhagen.
In Danish Standards Katrine is establishing and coordinating research projects in Danish Standards. She is also representing Danish Standards in the European working groups and projects under the auspices of CEN/CENELEC regarding research. At a national level she is managing the national contact point (NCP) for research in standardization.
Danish Standards is the national standardization organisation in Denmark working with formal standardization, and therefore the national member of Danish member of CEN, CENELEC, ISO and IEC.

Niels Madelung
Chief consultant
Niels Madelung is chief consultant at the Consulting Department at Danish Standards. He works with business development in the areas of service and product development, quality management, process optimization, risk management and business continuity. Niels is the author of the Danish book "Glidikkeibananskrællen" about risk management.

He is specialized in business and operational improvements, Lean Management, SixSigma (Yellow Belt), Quality Management, Risk Management, Business Continuity and he is an ISO standardization specialist with particular emphasis onISO 9001, ISO 26000, ISO 27001, ISO 27002, ISO 29100 and ISO 31000.

Niels is an expert in ISO/IEC SC 38, Cloud Computing, co-editor on revision of ISO/IEC 27002 Information Security Guideline and cofounder of The Danish Council for Greater IT-security.

# Business Impacts of International Standards for Information Security Management. Lessons from Case Companies

Robert M. van Wessel[1] and Henk J. de Vries[2]

[1]*Rotterdam School of Management, Erasmus University Department of Management of Technology and Innovation P.O. Box 1738, Room T10-42 3000 DR Rotterdam The Netherlands Phone: +31-6-28257307 Fax: +31-10-4089014*
*e-mail: rwessel@rsm.nl*
[2]*Rotterdam School of Management, Erasmus University Department of Management of Technology and Innovation P.O. Box 1738, Room T10-42 3000 DR Rotterdam The Netherlands Phone: +31-10-4082002 Fax: +31-10-4089014*
*e-mail: hvries@rsm.nl*

Received 18 January 2013; Accepted 14 May 2013

## Abstract

This paper describes the business impact of two international standards for information security management: ISO/IEC 27001 and ISO/IEC 27002. Six company cases show that companies had different reasons for wanting to implement these standards, but that they achieved most of their objectives. Benefits include improved service quality, higher customer satisfaction, and in some cases, new business opportunities. A number of common success factors ensure the objectives can be achieved, and financial and non-financial benefits can indeed be obtained. The lessons learnt from these cases can help other companies to also reap such benefits.

**Keywords:** Information security, ISO/IEC 27001, case study, standardization, business impact.

# 1 Introduction

## 1.1 Background

In today's digital age, information has become one of the most important assets of organizations and society as a whole. Information such as credit card details, medical records, and strategic business plans is increasingly processed and stored electronically and transmitted across the Internet. Properly protecting information is vital for competitive, legal and reputational reasons. Governments and businesses face considerable risks if information is not available, not reliable or unintentionally disclosed. 'Wikileaks' demonstrated that this has a major impact on individuals, organizations and society. Information security management deals with these threats, by assessing business impact, analysing vulnerabilities and applying suitable controls to achieve a balance between security, costs, usability and other business requirements.

## 1.2 Standards for Information Security Management

Systematic information security management needs accepted and reliable approaches. These have been developed and are laid down in standards. Different standards are available, but the most common and generally accepted framework by information security professionals is the standard ISO/IEC 27001 [1, 2] which is part of the ISO/IEC 27000-series of standards for information security management systems. These standards are published jointly by the International Organization for Standardization (ISO) and the International Electrotechnical Commission (IEC). They are intended to assist organizations of all types and sizes to implement and operate an information security management system (ISMS). The series provides good practices on information security management, risks and controls, and is similar in design to management systems for quality, environmental and IT service management (ISO 9000, ISO 14000 and ISO/IEC 20000 series respectively). ISO/IEC 27001 [2] specifies the requirements for establishing, implementing, operating, monitoring, reviewing, maintaining and improving a formalized ISMS within the context of the organization's overall business risks. The directly related standard ISO/IEC 27002 [3] provides a list of commonly accepted control objectives and best practice controls to be used as implementation guidance. These two standards originate from British Standard BS 7799, published in 1995.

At the end of 2010, at least 15,625 ISO/IEC 27001:2005 certificates had been issued in 117 countries. Most certificates were issued in Japan, India and the United Kingdom, with the highest growth in Japan, China and the Czech Republic. Figure 1 shows the growth in number of certificates [4].

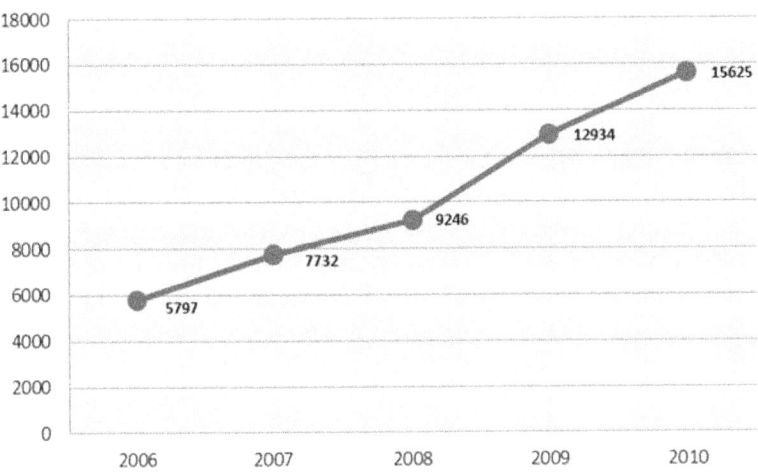

Figure 1 Number of ISO/IEC 27001 certificates [4].

## 1.3 Lack of Research in this Domain

With the growing importance of information security for business and society, an increased interest in this topic is found in academic literature [5]. Most contributions can be found in the fields of computer science and economics. A number of studies try to quantify the economic impact of information security, and some deal with information security management systems. Few address implementation [6] and impact [7, 8], and hardly any combine information security, organizational performance and the use of standards for information security management. In [9] a procedure is proposed to select the optimal investment of the required security technology based on the quantified value of an information system and capital investment appraisal techniques, whereas [10] proposes a Balanced Scorecard approach.

## 1.4 Research Approach

Due to this lack of research, BSI British Standards and the Chair of Standardization of the Rotterdam School of Management, Erasmus University initiated a common research project on the business impact ISO/IEC 27001 and ISO/IEC 27002 on companies. It provides evidence of positive impacts, and relates impacts to the way companies adopt these standards, so that other organizations can learn to also reap the benefits.

A conceptual model was used to assess the financial and non-financial impact for organizations [11]. The model includes the management and governance of the selection, implementation and use of the standard, and measuring its impact. Management of the standard is defined as the decision-making efforts associated with planning, organizing, controlling, and directing the selection, implementation and use of the standard in the organization. Governance of a standard is defined as specifying the decision rights and accountability framework to encourage desirable behavior in the selection, implementation and use of the standard in the organization. Impact is measured according to the Balanced Scorecard perspectives [12]. We use the Balanced Scorecard as it is a well-established method to measure performance, taking into account both tangibles and intangibles [13]. It includes the firm's current operating performance and its future performance drivers by measuring and tracking business performance from four perspectives: financial, customer, internal / business process, and learning & growth.

In-depth case study research was carried out at two companies in the UK and four companies in the Netherlands (see Table 1). Interviews with staff lasted between one and four hours.

All companies implemented relevant control objectives and best practice controls as specified in ISO/IEC 27002. Two companies acquired company-wide ISO/IEC 27001 certification, whereas four obtained certification for specific departments.

## 2 Results

The case companies adopted the standards for internal and external reasons. The most important internal ones were: 1) increasing the quality of services offered, 2) reducing the costs of security operations through standardized procedures and technical implementations to be applied in projects, and, 3) improving the company's risk profile. The most important external reasons were: 1) meeting customer requirements, 2) complying with legal requirements, and 3) improving brand and reputation.

ISO/IEC 27001/2 adoption resulted in a number of business benefits. Most companies reached most of their initial objectives and two achieved all of their objectives. One company accomplished only half of them. The primary cause

Table 1 Profile of the case companies that adopted ISO/IEC 27001/2.

| Case | Sector | Type | Country | Number of interviewees | Size | Certification scope |
|---|---|---|---|---|---|---|
| 1 | Financial Services | Commercial | NL | 5 | Very large | Some departments |
| 2 | Computing Services | Commercial | NL | 1 | Very large | Some departments |
| 3 | Financial Services | Commercial | NL | 2 | Very large | Some departments |
| 4 | Telecommunication | Commercial | NL | 2 | Very large | Some departments |
| 5 | Purchasing | Public | UK | 1 | Large | Enterprise-wide |
| 6 | Software Development and Maintenance | Commercial | UK | 2 | SME | Enterprise-wide |

was that it did not create standardized implementation (security) guidelines for projects. Each new project had to start from scratch, negatively impacting costs and throughput times.

In the following sections, we give a systematic overview of the impact on the business performance using the Balanced Scorecard perspectives.

## 2.1 Financial Perspective

The standards brought financial benefits, although the companies could not produce exact figures. Certain IT development costs increased (e.g., more project risk assessments) whereas other costs decreased, as less rework was required (e.g. no fixes due to lack of security). Operational losses fell but some operational costs rose (e.g., more vulnerability checks).

The majority of costs were incurred during the implementation of ISO/IEC 27001 or 27002. These included consultancy and certification costs. During the use phase, investments were made to guarantee secure information systems, and to develop staff training and induction programs for information security awareness. The standards resulted in better IT systems, and for some case companies, ISO/IEC 27001 certification resulted in new business opportunities and thus a positive financial impact.

## 2.2 Customer Perspective

The standards led to increased customer satisfaction. ISO/IEC 27001/2 certification provides reassurance to customers that the company's security status meets internationally accepted criteria and demonstrates credibility and trust that customer data is protected properly. Companies became more aware of potential risks and the standard leads to better understanding and cooperation between business-driven departments and the IT department. Of course, the customers should not be bothered with security implementations of the ISMS but in our cases they could see the difference: customer satisfaction on IT delivery and support improved.

Although certification was not a prerequisite for corporate clients of the telecommunication company (case 4), they responded positively to its ISO/IEC 27001 certificate. For two companies, certification was a prerequisite. One provided white labeling services (case 1). The other required certification because its competitors were already certified (case 2). One company (case 6) reported that it had had no major security incidents since certification, which contributed to customer confidence.

## 2.3 Internal Perspective

All companies achieved better quality of IT services and a reduced risk level. Authorization, IT asset management and change management improved. ISO/IEC 27001/2 certification increased awareness of business continuity, reduced the number of ad hoc activities, and led to a structural approach to resolve incidents. In a number of cases, availability of information systems increased. Little impact was found on time to market new services and time to develop and support IT.

Generally, staff were mildly positive about the standards. They became more aware of potential information security risks (threats and vulnerabilities). IT staff, however, were less positive as they considered many control measures to be bureaucratic, leading to an increase in workload. Staff opinions about the complexity and attractiveness of the standards depended on how the standards were implemented, ranging from a blindfold implementation of 133 control measures (negative) to selecting those that addressed high business risks first (positive). We also found that certification (enterprise-wide or at department level) improved the overall quality of the services offered and the risk profile of the company more than "just" being ISO/IEC 27001/2 compliant.

## 2.4 Learning & Growth Perspective

With relation to learning, a key advantage of the standards is that their broad acceptance in the information security profession leads to a common understanding of controls and vocabulary. Proven methods are used and staff do not need to reinvent the wheel. The standards can be considered as a framework for quality improvement in general, such as release management, version control and document management. Furthermore, the standards increase the level of quality thinking and risk awareness. With relation to growth, scalability of the IT infrastructure is easier because of the precautionary measures and these measures reduced the number of ad hoc security measures.

In some cases, ISO/IEC 27001 certification was a prerequisite for maintaining current business levels (cases 1, 2, 3 and 6), whereas for others it resulted in opportunities for new business. The telecommunication company (case 4) improved its competitive position through certification. For the most mature company (case 5), ISO/IEC 27001 was seen as business enabler. New jobs were created by developing new value-added secure service offerings.

Most of the case companies indicated positive effects during implementation and use due to familiarity with other quality management system standards, such as ISO 9001 and ISO 14001. Companies improved their market position by adopting an integrated set of such management system standards. One of the information security managers explained: *"Information security and green IT actually co-exist beautifully. Computer virtualization is a good example of this co-existence. It requires fewer resources as multiple computers systems are put logically on one machine, and security can be met more easily. Wherever you see a green policy, you see a security one that goes with it and vice versa. Why? Both are common sense, it is the right thing to do!"*

## 3 Analysis: Success Factors

Based on these case studies, we identified a number of success factors that ensure that the objectives and business benefits of ISO/IEC 27001/2 initiatives can be achieved.

### 3.1 Involvement of Business-Driven Departments

Probably the most important element of successful implementation is to ensure that ISO/IEC 27001/2 implementation is a business- rather than an IT-driven activity. Information security requirements should be determined by business-driven departments that can identify areas of weakness in risk assessments. Subsequently, these departments should implement controls that really add value, using a risk/benefit trade-off. Adequate governance and management of the ISMS is a key requirement.

In one of the companies (case 6), ISO/IEC 27001 was IT-driven as the IT director had initiated the project. At first, the steering committee met monthly. It took a while for management team to realize that without them actually proactively participating and filtering information down to their teams, the project was doomed. In another company (case 5) which had worked with ISO/IEC 27001 for many years, information security encompassed the entire company and was incorporated in everything the company did.

### 3.2 Senior Management Commitment

A second key element is to ensure commitment and endorsement for ISO/IEC 27001/2 at senior management levels. Although bottom up approaches may be successful to a certain extent, without the active support of senior management such initiatives will result in suboptimal results at best. For example, at the telecommunications company (case 4), the CFO initiated ISO/IEC 27001/2 adoption without top management endorsement. However, the chief information security officer ensured that full executive commitment was achieved, and the Board of Directors approved ISO/IEC 27002 implementation enterprise-wide. Again, adequate governance during selection and implementation determined the level of success.

### 3.3 Staff Involvement During Implementation

Another key element is to create awareness in all business-driven departments and the IT department, at both management and staff level. Prerequisite to success is continuous attention (e.g., weekly agenda item on implementation progress) by local MTs since these have to enable implementation. In the company cases, awareness was developed through activities such as workshops on risk assessments, asset classifications, business impact analyses, controls selection, and progress meetings

Based on the case company findings, the preferred implementation sequence was a combination of top-down activities (setting strategic objectives, policies and standards) and local bottom-up initiatives. A gradual implementation process was more successful than a 'big bang' approach, especially for the larger companies. For small companies, the big-bang approach was successful as well. Furthermore, a pragmatic and focused tactic (e.g., resolve major incidents, or requirements from regulators) was very effective. For example, one company started to tackle issues with logical access control by focusing on the five to ten most critical applications per business unit, and consequently the business-driven department became more involved. Furthermore, it turned out to be better not to speak in terms of specific controls (e.g. has control 10.4.2 "Controls against mobile code" been implemented?) but to discuss business implications of control measures. The following steps can be taken: 1) identify what should be protected, 2) determine the stakeholders, 3) perform risk analyses based on a standardised template, and 4) implement controls that really add value based on the specific risks (risk/benefit trade-off).

### 3.4 Continuous Improvement

Successfully completing an ISO/IEC 27002 implementation project and acquiring ISO/IEC 27001 certification are important milestones. However, ongoing improvement is vital to achieve and sustain long-term business benefits. We identified several key elements to success in this area. One is the importance of feedback from the user and project community about specified policies and standards. Policies and standards should be updated regularly as specified in Deming's well-known plan-do-check-act quality management cycle. The same holds true for implementation of (security) guidelines for projects to meet control measures. If the information security department creates standards or guidelines on how to implement measures such as access control, encryption or two-factor authentication, projects can get a head-start and project managers do not have to reinvent the wheel. For example, this will have a positive impact on costs, time to market, and interoperability.

Furthermore, disseminating information security policies, standards and best practices as part of security awareness is vital. Although information security awareness may exist among staff, when it comes to actual behavior, they often perceive information security controls as a hindrance to their normal routine. To maintain an acceptable level of security, information security awareness campaigns should be carried out twice a year and supported by website articles, emails, and warnings about hot topics. One of the companies (case 1) used desktop wallpapers as daily reminders of the importance of adequate information security behavior. Management should regularly assess staff satisfaction with the policies, standards and services offered by the information security department.

## 3.5 Clearly Defined Deviations Process

Another key element to the success of an ISO/IEC 27001/2 implementation project is the way a company deals with deviations from the prescribed controls. These controls should be mandatory but a number of case companies offered the possibility to temporarily deviate from these control measures, if accompanied by sound business rationale and mitigating factors. Other case companies did not allow any deviations but incorporated amended control measures into the ISMS if the current controls were no longer effective. Such decisions are all founded on proper risk-based analyses. The exception process differed per company. As long as the company considers this process adequately, the business-driven community will support information security. If not, it will find ways to circumvent the controls.

## 3.6 Experience with other Management System Standards

Another element of successful ISO/IEC 27001 implementation is familiarity with other management system standards. In one company (case 4), experience with ISO 9001 was an important success factor. Experience with the ISO 9001 quality management system resulted in 50% less time and effort spent on the implementation of ISO/IEC 27001 because improvement management, control mechanisms, context diagrams, and overviews of staff and systems were already available. One of the interviewees recommended companies to first implement ISO 9001 and only then ISO/IEC 27001. In another company at local management level (case 2), the general attitude towards using ISO/IEC 27001/2 was positively affected because of this same reason. During the ISO 9001 implementation project, management had initially opposed the quality management system. However, management had a much more positive attitude towards ISO/IEC 27001. This suggests a positive relationship between ISO 9001 and ISO/IEC 27001 adoption. The same is expected for similar management systems standards such as ISO 14001.

## 3.7 Governance and Management

Based on the case studies, we identified the following successful governance mechanisms. These support the findings of [11].

- Strategic and operational alignment of business-driven departments and the IT department in the selection process and operational phase. Business representatives should be more involved than IT staff. In all case companies, decisions about information security investment objectives and resource allocations were driven by the level of risk or based on regulatory requirements. The companies used a cost/benefit analysis to make these decisions.
- The integration of the controls, templates, etc. into existing processes and governance structures of the organisation. This includes, but is not limited to, strategy and year plans, project management, incident and change management, and IT service management tooling. In one of the organizations (case 4), information security project participation was first reactive, but the organization successfully leveraged its innovation structures and processes making it proactive.
- A convergence of functions in the security domain (information security, physical security, operational risk management, etc.). Functional embedding of information security differed per case company and most companies had a centralized set-up in either the IT or the corporate Risk Management departments. In the most mature organization, it was positioned just below board level.

We also identified the following successful management mechanisms:

- A positive attitude of management towards ISO/IEC 27001/2, such as commitment, priority setting and endorsement of the standard.Information security awareness programs. Awareness sessions for staff, evalua-

tions for management and workshops for both are essential to maintain and develop awareness.
- Information security staff should be attentive to changes in the organisation by anticipating changes in business needs, changes in management (structure), or new technology.
- An adequate waivers and dispensation process that allows companies to (temporally) deviate from the ISO/IEC 27001/2 standard and its related control measures.

## 4 Conclusions

The more organizations depend on information, the more important it is to ensure the confidentiality, integrity and availability of information systems. International standards have been developed to assist in this field. Standards provide a benchmark and good practice examples, guarantee that support is available (courses, consultancy, etc.), and certification signals achievements to external parties.

The international standards ISO/IEC 27001 and ISO/IEC 27002 are widely accepted by the information security profession. They are generally perceived as clear, pragmatic and logically structured, and contain proven concepts based on good practices. These standards accommodate a common language, improve awareness, communication and understanding of information security, increase customer satisfaction, enable the organization to offer more products; create business opportunities, and ISO/IEC 27001 provides the opportunity to acquire certification. Some organizations use ISO/IEC 27002 as a checklist that provides baseline measures. More experienced information security staff warn that this is the major pitfall as ISO/IEC 27002 has a vast number of control measures (133). The standard helps staff to think in terms of risk and risk management. Because of its risk based approach, a company only needs to mitigate those risks that are applicable to their business. Although not part of our case study, we anticipate the standards to be also beneficial for non-commercial organizations. Some of our case companies provide crucial services for society, such as public procurement, core financial services and telecommunications. In this way, the standards also contribute to a well-functioning society.

Our research contribution is threefold. First, we describe the business impact of ISO/IEC 27001/2. Many studies are available on the business impacts of other management standards, in particular ISO 9001 [14, 15, 16] and ISO 14001 [17]. We extend this by providing a study on the impact of a management standard in another domain. Second, this study shows that ISO/IEC 27001/2 adoption has led to overall improvements in service at both the technical and procedural level. Third, we have identified a number of good practices. By ensuring the success factors discussed earlier, the positive impact of ISO/IEC 27001/2 can be maximised. This study also confirms Taiwanese specific findings [6] that support from top management (section 3.2), awareness and education (section 3.4), and past experience with other standards (section 3.6), are important factors that influence the results of ISMS implementations. Other good practices include involvement of business-driven departments, commitment of senior management, an implementation process that combines top-down activities and local bottom-up initiatives, ongoing improvement of the ISMS, and a clear deviation process. In addition, experience with other management standards and effective governance and management mechanisms during the selection, implementation and use phases have a positive impact on a company's performance.

To conclude, companies can reap substantial benefits from the adoption of the international standards for information security management ISO/IEC 27001 and IEC 27002. The more the organisation depends on information systems, the more impact there is on performance. Our lessons on how to implement and use the standards may therefore be informative for many organizations.

## Acknowledgements

The authors thank the case companies for their willingness to share their experiences, and BSI and Netherlands Standardization Institute NEN for their support.

## References

[1] J. Backhouse, C.W. Hsu, L. Silva. Circuits of Power in creating de jure Standards: Shaping an International Information Systems Security Standard. *MIS Quarterly*. 30, 413-438, 2006.
[2] ISO/IEC, *ISO/IEC 27001 Information technology–Security techniques– Information security management systems–Requirements*. Geneva, Switzerland: International Organization for Standardization, and International Electrotechnical Commission, 2005.
[3] ISO/IEC, *ISO/IEC 27002 Information Technology—Code of Practice for Information Security Management*. Geneva, Switzerland: International Organization for Standardization, and International Electrotechnical Commission, 2005.

[4] ISO, "The ISO Survey of certifications 2010", Geneva, Switzerland: International Organization for Standardization, 2011.
[5] S. Ransbotham, S. Mita, "Choice and Chance: A Conceptual Model of Paths to Information Security Compromise", *Information Systems Research*, 20 (1), 121–139, 2009.
[6] C.Y. Ku, Y.W. Chang, D.C. Yen, "National information security policy and its implementation: A case study in Taiwan", *Telecommunications Policy*, 33(7): 371-384, 2009.
[7] A.G. Kotulic, J.G. Clark, "Why There Aren't More Information Security Research Studies", *Information & Management*, 41 (5), 597-607, 2004.
[8] J. L. Spears, "Institutionalizing Information Security Risk Management: A Multi-Method Empirical Study on the Effects of Regulation", Ph.D. Dissertation, Pennsylvania State University, 2007.
[9] R. Bojanc, B. Jerman-Blazic, "An economic modelling approach to information security risk management", *International Journal of Information Management,* 28 (5), 413–422, 2008.
[10] T. Herath, H. Herath, W.G. Bremser, "Balanced Scorecard Implementation of Security Strategies: A Framework for IT Security Performance Management", Information Systems Management, 27 (1), 72-81, 2010.
[11] R.M. van Wessel, "Toward Corporate IT Standardization Management. Frameworks and Solutions", Hershey, PA, USA: IGI Global, 2010.
[12] R.S. Kaplan, D.P. Norton, "The Balanced Scorecard - Measures that Drive Performance", *Harvard Business Review*, January-February 1992, 70 (1), 71-79, 1992.
[13] L. Willcocks, Information management. The evaluation of information systems investments, London: Chapman & Hall, 1995.
[14] E.L. Psomas, C.V. Fotopoulos, "A meta analysis of ISO 9001:2000 research - findings and future research proposals", *International Journal of Quality and Service Sciences*. 1, 128-144, 2009.
[15] B. Rusjan, M. Aliè, "Capitalising on ISO 9001 benefits for strategic results", *International Journal of Quality and Reliability Management*, 27, 756-778, 2010.
[16] P. Sampaio, P. Saraiva, A.G. Rodrigues, A.G., "ISO 9001 certification research: questions, answers and approaches", International Journal of Quality & Reliability Management. 26, 38-58, 2009.
[17] H.J. de Vries, D.K. Bayramoglu, T. van der Wiele (2012) "Business and environmental impact of ISO 14001", *International Journal of Quality & Reliability Management*, 29 (4), 425-435, 2012.

## Biography

**Robert M. van Wessel** holds a Master in Electrical Engineering from Twente University and a PhD in Business Administration from Tilburg University (Department of Information Systems and Management). He works as a Business Architect in the financial services industry and is associated with Rotterdam School of Management, Erasmus University. Robert's research interests relate to the interaction of Business and Information Technology, in particular Business Performance and the Value of IT, Enterprise Architecture, IT Governance, Portfolio Management, Information Security Management and IT Standardisation and Standards.

**Henk J. de Vries** is Associate Professor of Standardisation at the Rotterdam School of Management, Erasmus University, Department of Management of Technology and Innovation. His research and teaching focus on standardisation from a business point of view. Henk is President of the European Academy for Standardisation EURAS, Chair of the International Cooperation for Education about Standardization ICES, and Special Advisor to the International Federation of Standards Users IFAN. He is (co-)author of more than 300 publications in the field of standardisation. See http://www.rsm.nl/hdevries and http://www,rsm.nl/is.

# ICT Standards in South Eastern Europe (SEE) Education: Macedonian Case

Liljana Gavrilovska and Vladimir Atanasovski

*Faculty of Electrical Engineering and Information Technologies (FEEIT) – Skopje*

Received 30 March 2013; Accepted 14 May 2013

## Abstract

The Standards Education (SE) in the field of ICT gains increasing momentum worldwide. The strategic value of the ICT standards and their influence on the economy proves essential towards countries' development and their economic growth. This paper overviews the relevant current SE initiatives with a special emphasis on the South Eastern Europe (SEE) case and Macedonia. It discusses the level of ICT penetration, the recognition of the SE importance and the involvement of the relevant stakeholders in the SE curricula design on various education levels in Macedonia. Finally, the paper pinpoints the future directions towards transparent and harmonized SE.

**Keywords:** ICT, Standards Education (SE), South Eastern Europe (SEE), SE initiatives, ICT standardization, Macedonia's case.

## 1 Introduction

The Information and Communications Technology (ICT) represents a seamless convergence among the technologies dealing with information handling, software and communications, thus enabling the end-users with ubiquitous access, storage and manipulation of relevant information. The importance of ICT is rapidly increasing in the last decade promoting them into a pillar of the modern and well-developed societies. The ICT development is strongly affected and boosted by the *international standardization* that fosters harmonization on a worldwide level. Therefore, the **ICT standardization process** provides essential conditions and components for **transparent development of the overall society**.

The International Telecommunication Union (ITU) highlighted the importance of the ICT standardization through its Broadband Commission [1]. Its first country case studies published in 2012 showcase the links between the broadband connectivity and the UN Millennium Development Goals [2] in partnership with ITU. Figure 1 depicts the envisioned growth of the Internet penetration in developing and the Least Developed Countries (LDCs) worldwide. It is evident that the ambitious target of the ITU-UN partnership is to have 60% of the worldwide population online by 2015. The process of ICT standards harmonization proves to be crucial towards achieving this goal.

The rise of the ICT and its standardization is inevitably intertwined with the necessity for **Standards Education (SE)** in the field. The crucial point within is to understand the strategic value of the standards, the levers how this value is created and the effects in economic and public life. Therefore, a unified and harmonized worldwide SE can guarantee long-term ICT sustainability and its application in practice. It is of upmost importance that countries and universities work in this area providing transparent SE curricula ensuring education of new engineers and policy makers that can efficiently cope with future ICT challenges.

The Asian countries (e.g. Korea, Japan, China etc.) are currently leading the ICT SE process in the world [3]. They have established national bodies and action plans that treat the topic in a systematic manner starting from curricula development on all education levels (evenat primary school) through development of teaching materials and organization of conferences. This **top-down approach** proves to be the most effective one in terms of

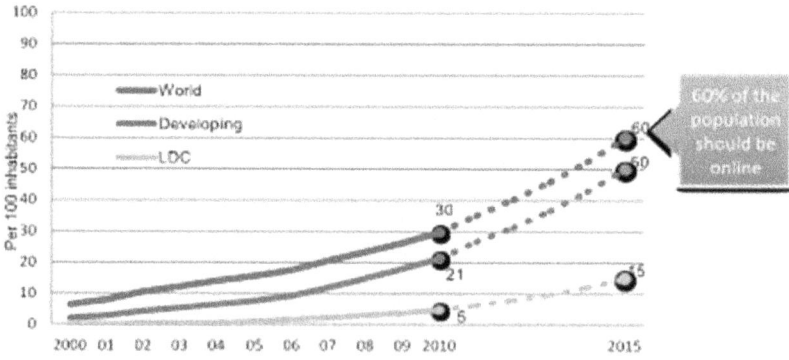

Figure 1 Broadband target 2015 (ITU-UN) [1].

market and industry needs for standards-literate professionals. Moreover, the Asian approach targets top students for SE education as the field is foreseen as a frontrunner for understanding the future ICT ecosystem, the interoperability of systems and the possible inconsistencies raising from disharmonized standards worldwide. Effectively, this ensures that the countries focusing on SE at this moment invest in their *future worldwide influence* and this must be embraced as a *roadmap by the developing countries as well*.

Unfortunately, Europe lags behind Asia in the SE field. The general consensus is that the SE at the academic level in Europe is inadequately based on the outcome of scientific research [3]. It means that Europe is very active in the area of *standardization research*, but *lacks effective SE*. Therefore, the European Academy for Standardization (EURAS) [4] was formed in order to stimulate the SE and promote the inclusion of education in standardization policies at national and European levels. At the same time, the South-Eastern Europe (SEE) countries are also vigorously trying to follow the European and the world trend in SE. Most of them realized the potentials of the ICT for the developing societies and focus on SE. In this manner, the **INA Academy** [5] tries to act as a catalyst for ICT development in SEE promoting diffusion of ICT services and SE in the field.

Republic of Macedonia is a developing SEE country that **already recognized ICT as a strategic area towards its future growth**. Relevant national bodies and universities have started a dialog on creation of SE curricula, which would fit the needs of modern ICT and will ensure that Macedonia keeps the momentum in the area with the developed countries. Also, the country participates in numerous international standardization bodies that foster the process of ICT SE initialization and harmonization.

This paper discusses the SE status worldwide, with a particular emphasis on the SEE case and Macedonia. Its focus is on the level of ICT penetration in the Macedonian society and the corresponding SE efforts and their harmonization. The paper is organized as follows. Section 2 provides general overview on the SE worldwide. Section 3 focuses on the Macedonian case elaborating the ICT status and standardization as well as the participation towards SE activities. Section 4 briefly describes the current university experiences and practices in Macedonia regarding the SE. Finally, section 5 concludes the paper.

## 2 Relevant International SE Initiatives

The importance of the ICT SE area emphasizes the potential negative consequences if SE is neglected. This can lead to serious setbacks and loss of market influence by the involved stakeholders. Therefore, the SE area is gaining momentum by a plethora of international organizations such as the ITU [1, 5], International Standards Organization (ISO) [6], International Electrotechnical Commission (IEC) [7], International Cooperation for Education about Standardization (ICES) [8], Institute of Electrical and Electronics Engineers (IEEE) [9] etc. Their SE related initiatives range from harmonization of the ICT standards to raising SE awareness via public events (e.g. workshops, seminars etc.) and creation of standard related curricula. This will ensure the leading strategic role of the ICT in the $21^{st}$ century.

The pioneering steps towards transparent SE worldwide were established in 2001 when IEC, ISO and ITU jointly established the **World Standards Cooperation (WSC)** [10]. The main idea behind was to strengthen and advance the voluntary consensus-based international standards systems of IEC, ISO and ITU. WSC also promotes the adoption and implementation of international consensus-based standards worldwide and resolves any outstanding issues regarding cooperation in the technical work of the three organizations. The WSC initiatives comprise public events, education and training.

On an international level, ICES is also strongly committed to facilitating SE by providing guidelines for curricula development and creation of education materials. ICES is a network of individuals and organizations whose mission is topromote education about standardization and improve its quality and attractiveness for all stakeholders.

The Asian countries, i.e. South Korea, Japan and China, currently hold the primate in SE worldwide. Korean Standards Association (KSA) [11] body ensures supply of adequate expertise and coordination of the university curricula on SE. The university SE curricula are broadly present in the countries academic education system with multidisciplinary one-semester courses accompanied by relevant materials supplied from Korean standards experts. Korea initiated University Education Promotion on Standardization (UEPS) program in 2004 establishing itself at the forerunning place in SE worldwide. Japan takes a similar and well-devised approach to SE investing 1M US dollars in the period 2005-2010 for identification and addressing of all SE-related issues [12]. Moreover, the focus of the Japanese SE is on the strategic aspects of standardization (e.g. business strategies, intellectual property rights, strategic management etc.) rather than on the technological ones. Indonesia also follows the Asian trend hosting a WSC Academic Day in 2012 where 28 universities participated in discussions on SE. Finally, China provides several curricula on SE on more than 30 universities targeting hands-on experience in standardization, e.g. industry internships. The Chinese National Institute for Standards (CNIS) [13] is monitoring the SE process in the country paying serious attention to research and development for its own business of standardization.

Europe recently started to promote initiatives in the SE area [14]. The most important aspect is the need for a coherent approach towards SE on a continental level. Europe acknowledges that the possible consequences of neglecting SE in Europe can be summarized as [14]:

- Reduction of the European influence in international standardization;
- Domination of non-European agents in international standardization and
- Europe becomes a follower rather than a leader in standardization issues.

As a result, every EU country must closely follow the SE initiatives in order to guarantee the SE transparency.Currently, European leader in SE is Britain [15] closely followed by Germany and France.

European efforts towards SE and harmonization are best reflected in the joint initiative of the European Committee for Standardization (CEN) [16], the European Committee for Electrotechnical Standardization (CENELEC) [17] and the European Telecommunications Standards Institute (ETSI), which formed a **Joint Working Group – Education about Standardization (JWG-EaS)** [18] with hopes to professionalize education in standardization and increase the number of people who have a fair and positive knowledge of standardization, its characteristics and its added value. This initiative is very important for SEE countries (including Macedonia), as most of them are ETSI and ITU members.The focus of JWG-EaS is aimed at [18]:

- Creation of a network of national members interested in education about standardization;
- Provisioning of a concerted policy on education about standardization in order to maximize the benefits amongst the National Standards Bodies in Europe and abroad;
- Gathering of best practice to convince governments and regions, academia, companies etc. of the value of education about standardization and propose appropriate actions;
- Establishing and maintenance of appropriate contact with other activities relevant to education about standards, for example EURAS [4] and ICES and
- Setting up a repository of tools and materials concerning education about standards and standardization.

As already mentioned, **EURAS** also governs relevant SE activities. EURAS is a German-based society that promotes research, education and publication in the field of standardization through organization of various dissemination events. The most prominent one is the annual EURAS Conference that gathers researchers, policy makers and company representatives interested in the field of SE. EURAS is devoted to stimulating SE on a European level in a more systematic and coherent manner. In practice, many European researchers weakly correlate their work with educational activities leaving Europe behind Asia in the field of SE. Therefore, EURAS is strongly committed to mapping the standardization research into relevant SE and corresponding possible academic contributions to standardization as well.

The SEE countries are trying to follow the European current trend on SE by promoting various initiatives mostly in the form of public dissemination (e.g. conferences, workshops etc.). The most serious organized contribution to SE in SEE is done by the INA Academy [5], which closely monitors the ICT developments and enables informational update in the areas of regulatory management, customized research, business modeling/forecasting and technology strategy. However, the SE is generally targeted towards senior managers from ministries, regulators and operators through various forms of trainings. Figure 2 shows the geographical focus of the INA Academy. It is evident that the Academy covers EU and non-EU members from SEE building an extensive network of high-profile contacts in national bodies and operators resulting in brainstorm meetings, intensifying of the ICT involvements, assistance in finding funding mechanisms, harmonizing the EC guidelines and standards and providing (partial form of) SE in the region.

The relevant SE activities on an international level must be carefully scrutinized and applied in a harmonized manner locally. This will ensure that the targets are met and that the SE truly serves its purpose towards transparent development of the overall society. The following section will provide a case study on ICT penetration, standardization and SE in Macedonia.

## 3 Case Study: Macedonia

As mentioned in the introduction, Republic of Macedonia is a developing SEE country recognized as a nation with strong commitment to connectivity as a driver of national growth [19]. Macedonia boasts an impressive broadband penetration rate of 32% on a national level with 100% company Internet connectivity. Moreover, the Internet access in schools and WiFi-based public Internet access is already rolled out with very high percentage of national

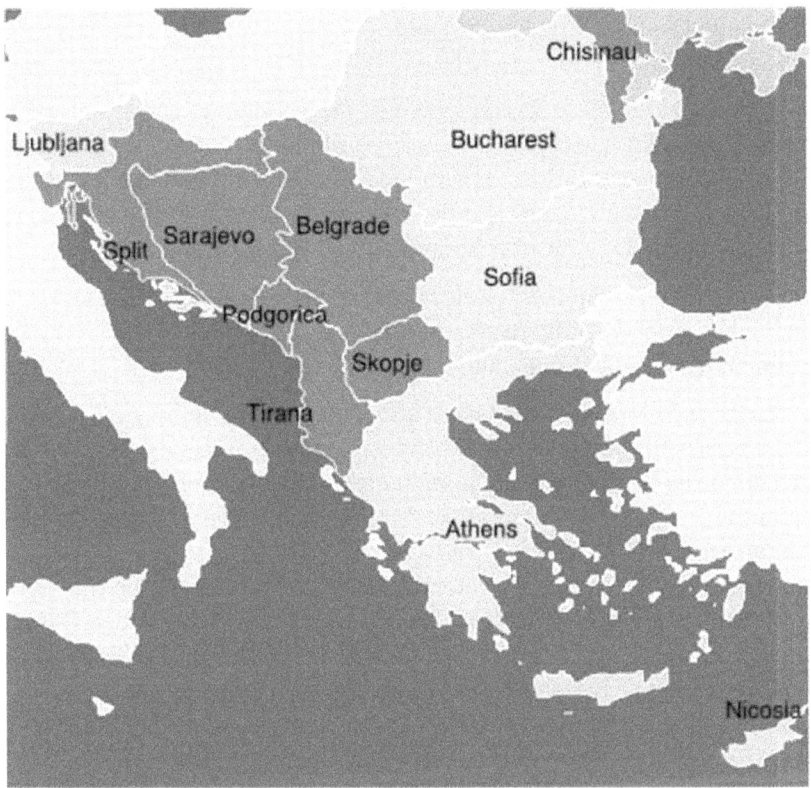

Figure 2 Geographic focus of the INA Academy (red colour indicates associated EU country, yellow colour indicates EU country member) [5].

coverage including remote areas. Macedonian schools offer one web-enabled computer for every 1.45 children. University students and academics can freely access knowledge and research resources via the academic network MARnet.

It is evident that Republic of Macedonia is a representative example of an ICT embracer. This section will provide details on the current ICT status in general in the country as well as on the ICT standardization and SE related activities.

### 3.1 ICT Status

The ICT sector is a very vibrant one in the country. It is **strategically led and backed by the Government** through two distinguished documents:

- Information Society Strategy [20] and
- National Broadband Strategy [21]

The *Information Society Strategy* defines the economic, social and political vision of the knowledge based society through ICT development and application in all spheres of life. Its aim is to foster creation of modern and efficient services for the citizens and the business community. The adoption of the ideas in the strategy by all relevant national stakeholders resulted in:

- Entirely liberalized market for electronic-communication services;
- Significant number of Internet users and
- Establishment of electronic public services.

The *National Broadband Strategy* is aimed at bridging the digital divide and providing broadband penetration comparable to the one in the developed European countries. The document complements the Government efforts to promote the ICT as a cornerstone of the Macedonian society reconstruction and development.

The adoption of the national strategies led to an increase in the Internet penetration in the country promoting the usage of ICT on a wider scale. Figure 3 depicts the Internet penetration in SEE [22] whereit is obvious that Macedonia represents a regional leader in the area.

The ICT responsibilities in the country were centralized through the creation of a Ministry for Information Society and Administration (MIOA) [23] that closely monitors all ICT related developments including the standardization and the SE. Also, MIOA is responsible for implementation of governmental ICT politics, thus contributing to the achievement of all objectives in the

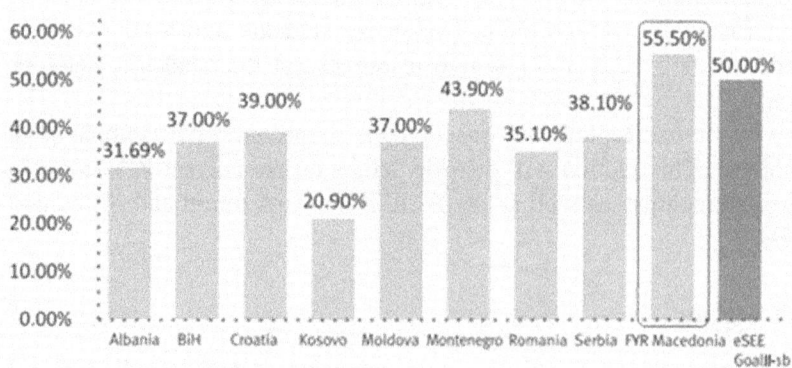

Figure 3 Internet penetration in SEE [22].

previously mentioned strategies. MIOA is in close cooperation with several national institutions such as the Agency for Electronic Communications (AEC) [24], the Standardization Institute of the Republic of Macedonia (ISRM) [25] and the Ministry of Transport and Communications (MTC) [26].

As a result of the joint, systematic and coherent approach by all ICT stakeholders, this area emerges as a strategic one in education and in the Macedonian society in general.

## 3.2 ICT Standardization

Republic of Macedonia participates in numerous international organizations targeting (or making use of) ICT standardization:

- International Telecommunication Union (ITU);
- European Telecommunication Network Operators' Association (ETNO);
- European Telecommunications Standardization Institute (ETSI);
- International Satellite Communications Organization (INTELSAT);
- European Satellite Communications Organization (EUTELSAT);
- International Satellite Organization (INMARSAT);
- European Broadcasting Union (EBU);
- European Conference of Postal and Telecommunications Administrations (CEPT) and
- World Trade Organization (WTO).

The international activity of the country in these organizations fosters the ICT standardization process ensuring accurate and up-to-date ICT development. The international ICT standardization efforts are channelized through the most important national ICT pillars, i.e.:

- MASIT - ICT Chamber of Commerce;
- Economic Chamber of Macedonia / IT Association - Macedonian Association of the IT Companies;
- Macedonian Chambers of Commerce / ICT Chamber;
- Macedonian Academy of Sciences and Arts;
- Macedonian e-Society Association (MESA);
- e-Gov Project and
- Foundation for sustainable ICT solutions "Metamorphosis".

The focal Macedonian standardization body is the **Standardization Institute of the Republic of Macedonia (ISRM)** [25]. As of June 2012, ISRM is a full-fledged member of CEN and CENELEC, thus closely monitoring the international ICT standardization and its application on a national Macedonian level. The accession of ISRM in CEN and CENELEC is viewed as a very positive development, not only for the European standardization system, but also for the economy of the Balkan region and the whole of SEE. It means that the Republic of Macedonia can become fully integrated into the European single market.

ISRM's ongoing activities target active participation in relevant ICT entities such as the EC COST [27], the EC FP7 [28] programme, the NATO SfP programme [29]etc. promoting ISRM into a contributor to international ICT standardization. Moreover, ISRM's activities also focus on cooperation and facilitation of transition processes towards European initiatives (e.g. transition to digital terrestrial television etc.). Finally, ISRM is responsible for improvement of knowledge and awareness on ICT standards among industry, SMEs and consumers paving the way for SE curricula development in the country.

## 3.3 Participation in SE Activities

Republic of Macedonia acknowledges the importance of SE and its potential impact towards future ICT-oriented developed societies. The most important lessons to be learned and applied from the worldwide experiences can be summarized as:

- Introduction of SE on all educational levels in a systematic, top-down, approach, i.e. providing teaching materials and repositories;
- Selection of high-quality students for SE;
- Organization of public dissemination events on ICT standards and their importance in the developing societies and
- Active participation in national and international standardization organizations.

Currently, the participation in SE activities in Macedonia is organized on educational (secondary and higher education) level and national regulatory level.This subsection will provide more details on the ongoing initiatives.

### 3.3.1 SE in High Schools and Universities

The SE in high schools and universities is mostly channelized through the ICT curricula on secondary and higher levels. There are numerous ICT educational centres scattered throughout the country, Figure 4, offering education in various ICT areas. Even though the SE is inherently present in every ICT curricula as a fundamental part, the number of dedicated ICT SE programmes, especially in the universities, is quite low.

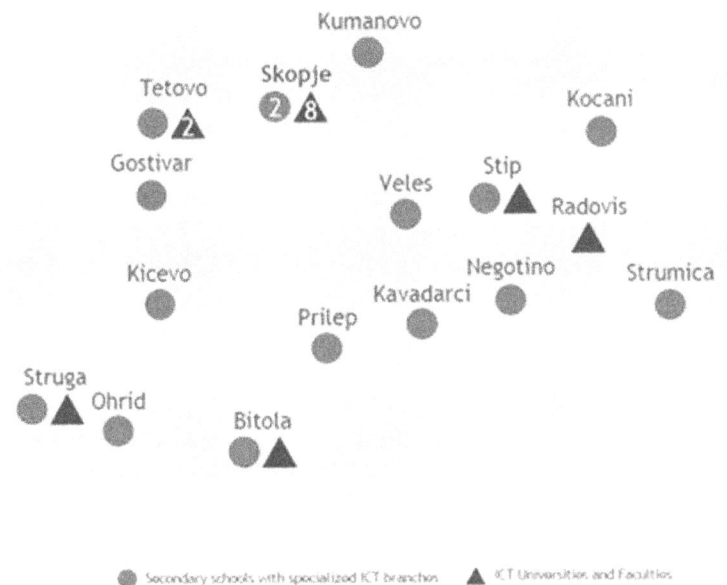

Figure 4  Distribution of ICT educational centres in Macedonia [31].

Almost all higher education institutions and universities in the country offer an ICT curriculum. The most versatile programme along with dedicated ICT SE initiatives is being offered on the largest and oldest university in the country, i.e. **Ss. Cyril and Methodius University in Skopje (UKIM)** [30]. More details on the university practices in ICT SE in Macedonia are given in section 4.

### 3.3.2 National SE Efforts

According to Article 73 of the Stabilization and Association Agreement (SAA) between the Republic of Macedonia and the European community and its countries (April 2004), Republic of Macedonia undertakes all necessary measures in order to speed up the development of the standardization as one of the pillars in the quality infrastructure, to support the participation in the work of the European standardization bodies (e.g. CEN, CENELEC, ETSI etc.) as well as to encourage the use of Community technical regulations and European procedures for standards, testing and conformity assessment [25]. The *Law on Standardization* ("Official Gazette of the Republic of Macedonia",

No. 54/02) regulates national standardization efforts. Republic of Macedonia acknowledges the importance of ICT, clean energy, energy efficiency and robotics as the most important areas towards standards harmonization [20, 25].

The main Macedonian national body dealing with the standardization in general (and therefore with ICT standardization also) is the ISRM [25]. The main strategic goal of ISRM is the harmonization of international standards locally, thus ensuring the fulfillment of the necessary prerequisites in the standardization area for Macedonia's European Union membership. ISRM actively participates in the work of European standardization bodies, provides ICT resources, adopts European and withdraws conflicting national standards, implements notification and standstill procedures and guarantees for protecting the rights of CEN and CENELEC publications. It also aims to develop a **standardization strategy** to ensure and encourage the involvement of all relevant stakeholders.

ISRM's SE efforts are disseminated through various public events such as workshops and seminars. There was a joint ISO/ISRM Awareness-Raising Seminar in 2009 [32] that gathered 29 participants from production, **educational institutions** and governmental bodies to discuss the need and the benefits of harmonized standardization and SE. Also, ISRM organizes workshops on its participation in the CEN/CENELEC technical committees in order to promote the SE work in the area.

Besides ISRM, there are also numerous national SE efforts lately. The national regulatory body AEC is raising awareness on broadband and ICT standardizationthrough organizing IRC conferences [33]. The IRC conferences serve as a platform for exchange of ideas among regulators, operators and vendors in SEE emphasizing the importance of standards literacy for future interoperability and society development. It presents and highlights the latest standards such as LTE-Advanced, spectrum regulation future strategies, broadband and multimedia protocols and standards. Also,ITU is locally present through its *Centre of Excellence (CoE)* at the Faculty of Electrical Engineering and Information Technologies (FEEIT), Ss. Cyril and Methodius University in Skopje (UKIM) that regularly offers e-learning courses to interested parties, e.g. regulators and vendors. These courses target the latest ICT technologies and their standardization on a global scale. *FEEIT/UKIM* additionally offers various SE related courses in its curricula for the undergraduate and the graduate students. Furthermore, FEEIT/UKIM regularly gives courses to broad professional auditorium (form of continued education) on new and evolving technologies and standards such as UMTS, LTE, IPv6, VoIP etc. Moreover, various vendors present in the country offer the possibility to

ICT professionals to obtain a standard-based certificate for proficiency in a certain ICT technology, e.g. *Cisco academia* on networking etc.Finally, many universities and companies regularly take part in FP7 related activities on research, development and harmonization in the SE field.

It is evident that the SE efforts are gaining momentum and that all relevant stakeholders (academia, vendors, regulators) are actively participating. However, all these efforts are sporadic and isolated and there is a clear need for systematic approach towards SE. A possible solution is to embrace the SE concept on a university level and offer specially created programs for SE on all academic levels. The following section will discuss in more details the current status of the SE university practices in the Republic of Macedonia.

## 4 University Practices in ICT SE in Macedonia

The initial efforts to provide relevant SE in the field of modern ICT emerged in 2004 at FEEIT/UKIM for the students majoring "Telecommunications" on undergraduate level and "Wireless and mobile communications" and "Communications and Information Technology" on graduate level. In 2011, the undergraduate curriculum was upgraded and enhanced to better suit the latest ICT initiatives, thus there was a change of the name into "Telecommunications and Information Engineering". Students study a handful of courses based on relevant ICT standards such as the IEEE 802 family of standards, 3GPP, optical networks, DOCSIS, multimedia etc. There are also **specific ICT SE related courses** such as "*Standardization and regulations in telecommunications*" (offered on a Dipl.-Ing. level) and "*Business management in telecommunications*" (offered on a MSc level).

They introduce the students to the activity domains of the most important standardization bodies (e.g ETIS, ITU, 3GPP, IEEE) and the most relevant standards, highlighting the importance of standards towards shaping and boosting the ICT developments and competitiveness.

Additionally, the curricula are continuously updated to accommodate the new technologies, e.g. LTE, LTE-Advanced, DVB-T etc. As a result, students majoring telecommunications on undergraduate and graduate level are trained to understand the ICT standardization process are familiarized with practically all relevant ICT standards today and are well aware of the benefits of harmonized SE on a national and international level.

In order to further extend its involvement in SE and be in line with the international corresponding efforts, FEEIT/UKIM created a specific curriculum termed "**Regulation in energetics, electronic communications and traffic**" on a graduate level in February 2013. This is the **first organized effort** to provide a completely *SE devoted curriculum on an academic level*. The choice of the courses and their contents was carefully designed in accordance with the latest industry developments and industry needs for harmonized SE. The curriculum caters for all relevant ICT standards for electronic communications, both fixed and wireless.

Additionally, there is an effort between the national AEC and UKIM to jointly develop an *ICT SE curriculum for the domain of spectrum usage and novel wireless services and initiatives*. This is still an ongoing issue.

It is clear that the Macedonian academic efforts to provide SE are following the international trends showcasing the benefits of harmonized SE for the overall development of the modern day ICT-based society in general.

## 5 Conclusions

The ICT in general is widely recognized as a driver towards modern society development. ICT is emerging as an inevitable part of all aspects of modern living, thus there is a clear need of understanding and embracing the ICT concept in order to fully use its potential. This is where the SE paradigm raises emphasizing the potential gap between the ICT developments and the society's potential to transparently follow them. Therefore, there must be a **focused and harmonized SE effort on international and national levels** enabling the bridging of the digital divide and creating an ICT literate future society.

There are several opportunities for provisioning SE towards fulfilling its goals. All relevant stakeholders (i.e. academia, vendors and regulators) from the public and the private sector should be involved in the process by organizing dissemination events, raising awareness through seminars and workshops etc. This would ensure that a systematic and coherent approach is held guaranteeing successful SE.

It is clear that the **academia can play a key role in the SE process** by introducing relevant curricula on all levels of studies. However, the academia must work in close cooperation with other interested parties, specially the standardization responsible bodies in order to create the most suitable market compatible curricula. These *synergies* will faster the developments towards a comprehensive ICT-based society.

This paper focused on the SEE initiatives for SE and especially on the case study of the Republic of Macedonia. Evidently, the region undertakes a number of actions towards SE awareness and active participation in the ICT standardization harmonization area. It should be noted that the ICT area is very vibrant and needs a pro-active approach for efficient SE and ICT incorporation in all areas of modern day living.

## References

[1] International Telecommunication Union (ITU), Broadband Commission for Digital Development. Information available at: http://www.broadbandcommission.org/.
[2] United Nations Millennium Development Goals. Information available at: http://www.un.org/millenniumgoals/.
[3] Hesser, W., and H. J. de Vries, "Academic Standardisation in Europe," *White paper*, EURAS, August 2011.
[4] European Academy for Standardization e.V. (EURAS). Information available at: http://www.euras.org/.
[5] ITU INA Academy Center of Excellence. Information available at: http://academy.itu.int.
[6] International Organization for Standardization (ISO). Information available at: http://www.iso.org/iso/home.html.
[7] International Electrotechnical Commission (IEC). Information available at: http://www.iec.ch/.
[8] International Cooperation for Education about Standardization (ICES). Information available at: http://www.standards-education.org/.
[9] Institute of Electrical and Electronics Engineers (IEEE). Information available at: http://www.ieee.org/index.html.
[10] World Standards Cooperation. Information available at: http://www.worldstandardscooperation.org/.
[11] Korean Standards Association (KSA). Information available at: http://www.ksa.or.kr/eng/.
[12] Tanaka, M., "Cooperation between standards bodies, academia and professionals from companies in Japan," *ICES workshop and WSC Academic Day 2011*, Hangzhou, China, 2011.
[13] China National Institute of Standardization (CNIS). Information available at: http://en.cnis.gov.cn/.
[14] Hesser, W., and A. Czaya, "Proposals for a coherent approach to standardisation education in Europe," *Educating a New Generation of ICT Standards Professionals, EU Commission Workshop*, Brussels, Belgium, Nov. 2009.
[15] BSI Education. Information available at: http://www.bsieducation.org/Education/default.php.
[16] European Committee for Standardization (CEN). Information available at: https://www.cen.eu/cen/pages/default.aspx.
[17] European Committee for Electrotechnical Standardization (CENELEC). Information available at: http://www.cenelec.eu/.

[18] CEN-CENELEC-ETSI Joint Working Group on Education about Standardization (JWG-EaS). Information available at:http://www.cencenelec.eu/standards/Education/JointWorkingGroup/Pages/default.aspx.
[19] ITU Broadband Commission, *Strategies for the promotion of broadband services and infrastructure: a case study of TFYR Macedonia*, First Series Country Case Studies, Geneva, Switzerland, May 2012. Information available at: http://www.itu.int/ITU-D/treg/broadband/BB_MDG_Macedonia_BBCOM.pdf (Last accessed: Mar. 26, 2013).
[20] National Strategy for Information Society Development of the Republic of Macedonia: Action Plan, Government of the Republic of Macedonia, Commission for Information Technology, Skopje, Macedonia, April2005. Information available at: http://www.mio.gov.mk/files/pdf/en/Strategija_i_akcionen_plan.pdf (Last accessed: Mar. 26, 2013).
[21] National Strategy for the development of Electronic Communications with Information Technologies: Strategic Directions, Republic of Macedonia, Ministry of Transport and Communications. Information available at: http://www.mio.gov.mk/files/pdf/en/NSEKIT_English-Parlament%20_2.pdf (Last Accessed: Mar. 26, 2013).
[22] United Nations Development Program (UNDP),*eGovernance and ICT Usage Report for South East Europe – $2^{nd}$ edition*, Sarajevo, Bosnia and Herzegovina, 2010.
[23] Ministry of Information Society and Administration (MIOA) of the Republic of Macedonia. Information available at: http://www.mio.gov.mk/.
[24] Agency for Electronic Communications (AEC) of the Republic of Macedonia. Information available at: http://www.aec.mk/.
[25] Standardization Institute of the Republic of Macedonia (ISRM). Information available at: http://www.isrm.gov.mk/.
[26] Ministry for Transport and Communications (MTC) of the Republic of Macedonia. Information available at: http://www.mtc.gov.mk/.
[27] Intergovernmental Framework for European Cooperation in Science and Technology (COST). Information available at: http://www.cost.eu/.
[28] European Commission Seventh Framework Programme (FP7). Information available at: http://cordis.europa.eu/fp7/home_en.html.
[29] NATO Science for Peace (SfP) Programme. Information available at: http://www.nato.int/science/.
[30] Ss. Cyril and Methodius University in Skopje (UKIM). Information available at: http://www.ukim.edu.mk/.
[31] Ministry of Education and Science of the Republic of Macedonia (MON). Information available at: http://www.mon.gov.mk/.
[32] Joint ISO/ISRM Raising-Awareness Seminar. Sep. 2009.
[33] International Regulatory Conference 2012 (organized by AEC). Information available at: http://www.aek.mk/index.php?option=com_content&view=article&id=552&Itemid=230&lang=en.

## Biography

Prof. Liljana Gavrilovska currently holds the position of full professor and Head of the Institute of Telecommunications at the Faculty of Electrical Engineering and Information Technologies, Ss Cyril and Methodius University in Skopje. She is also Head of the Center for Wireless and Mobile Communications (CWMC) working in the area of telecommunication networks and wireless and mobile communications. She has received her B.Sc, M.Sc and Ph.D. from Ss Cyril and Methodius University in Skopje, University of Belgrade and Ss Cyril and Methodius University in Skopje, respectively. Prof. Gavrilovska participated in numerous EU funded projects such as ASAP, PACWOMAN, MAGNET, MAGNET Beyond, ARAGORN, ProSense, FARAMIR, QUASAR and ACROPOLIS, NATO funded projects such as RIWCoS and ORCA and several domestic research and applicative projects. In 2012 Prof. Gavrilovska got a Scientist of the year UKIM award. Her major research interest is concentrated on cognitive radio networks, future mobile systems, wireless and personal area networks, cross-layer optimizations, broadband wireless access technologies, ad hoc networking, traffic analysis and heterogeneous wireless networks.

Prof. Gavrilovska is author/co-author of more than 150 research journal and conference publications and technical papers and several books. She is a senior member of IEEE.

Dr. Vladimir Atanasovski currently holds the position of assistant professor at the Institute of Telecommunications at the Faculty of Electrical Engineering and Information Technologies, Ss Cyril and Methodius University in Skopje. He has received his B.Sc, M.Sc and Ph.D. from Ss Cyril and Methodius University in Skopje, in 2004, 2006 and 2010, respectively. Dr. Atanasovski participated in numerous EU funded projects such as PACWOMAN, MAGNET, ARAGORN, ProSense, FARAMIR, QUASAR and ACROPOLIS, NATO funded projects such as RIWCoS and ORCA and several domestic research and applicative projects. Dr. Atanasovski is an author/co-author of more than 90 research journal and conference publications and technical papers. His major research interests lie in the areas of cognitive radio networks, resource management for heterogeneous wireless networks, traffic analysis and modeling, cross-layer optimizations, ad-hoc networking and nanonetworks.

# Global Standardization Education Program Collaborated by Osaka Univ. and MJIIT, UTM

Hiroshi Nakanishi[1], Rozhan Othman[2], Kunio Igusa[2], and Shozo Komaki[2]

[1] Osaka University, nakanishi@idiscp.osaka-u.ac.jp
[2] Malaysia-Japan International Institute of Technology, Universiti Teknologi Malaysia, rozhan@ic.utm.my, kunigusa@gmail.com, komaki@ic.utm.my

Received 27 March 2013; Accepted 14 May 2013

## Abstract

In Osaka University, the Education on Global Standardization Program (Univ-EoGSz) for post-graduate students is offered by the Interdisciplinary Center of Osaka University and extensively developed since 2009. The Malaysia-Japan International Institute of Technology (MJIIT), Universiti Teknologi Malaysia (UTM) is now going to setup the courses of standardization education as a sub program in collaboration with Osaka University for Malaysian academics and university students. In this article, the objectives and frameworks of this global standardization education program will be evaluated, especially on the existing Osaka Univ. program and MJIIT program. The general issues on standardization education and the results of case studies on the current trends of global standardization are discussed. The contents of Osaka Model, which MJIIT is now adopting, are slightly different from the conventional educations regarding the introduction of standards, which was architected for the general local schools and companies. It is expected that the article will give a valuable example for understanding on the future direction of standardization educations and advance further its activities in academic and/or university on the issue. And by promoting this education, we hope the conceptual paradigm regarding standardization will be shifted from the just simple *international* standardization, as reflecting international expansion of the intellectual property right (IPR) based national standard, to *global* standardization, as commonly acceptable one beyond IPR. And it focuses more enhancing the innovativeness, entrepreneurial and global standardization activities as well.

**Keywords:** Education on Global Standardization, Generic, Academic, University.

## 1 Introduction

There has been considerable effort in expanding standards education programs by the various local standardization bodies. This includes programs offered by national standardization bodies, industry supported consortium/forums and internal organization of industries. In the past, these programs are provided to companies and local schools, and were mainly focused on the educating students and trainees of the established or revised standards and their certification process. This includes standards set by national and/or international standardization bodies.

Development of such standards were mainly generated based on the interests of limited number of nations or companies, which have already developed and locally/nationally implemented products based on their intellectual property rights (IPR). These standards are referred to as international standards (IS). In recent years, the World Trade Organization (WTO) issued technical barriers of trade (TBT) and fair, reasonable and non-discriminatory licensing (FRAND/RAND) and this ruling is applied all over the world. Under this ruling, developed standards are referred to as global standards (GS) and not international standards, because the role of IPR is given less emphasis and national standards have less importance. So the global standardization (GSz) activitiesis become more important than IPR and national standards (NS) setup. Under these situations, standardization activities meet paradigm shift from IPR and NS oriented international standardization to global standardization oriented activities. After paradigm shift, ongoing standardization with development is required and global discussions among global interested parties are inevitable. This is called as Global Standardization (GSz), not international standardization (ISz).

Education on standard setting to companies and schools in Japan also reflects this paradigm shift. The earlier standard education has been mainly focused on the propagating the established standards and certifications. This had in the past been referred to as education on international standard for industry (Industry-EoIS). After the paradigm shift, education on global standardization for academic members and university students becomes more important to fit for the ongoing standardization with development. We call this as Education on Global Standardization for University (Univ-EoGSz), and set up the program and courses under the collaboration with MJIIT, UTM and Osaka University.

In this article, section 2 describes the importance and objectives for standardization education in/for faculty members and university students. Section 3 introduces the Osaka University's global standardization program for post-graduate students, which contains several courses relating with global standardization. Section 4 introduces the global standardization education sub-program, which is now planning to introduce in MJITT, UTM through their collaboration with Osaka Univ. Section 5 introduces the relationship between Malaysian national standardization bodies and global standardization bodies, and also our planning sub-program. This article is not a technical paper, however it is focused on education on global standardization in/for university (Univ-EoGSz),and it will become important activity after the paradigm shift in standardization.

## 2 Objectives for Standardization Education in University

### 2.1 Paradigm Shifts in Standardization

The history of standardization can be traced back to the ancient civilizations of Babylon and early Egypt. The earliest standards used were the physical standards for weights and measures. By the time of the Industrial Revolution in the early nineteenth century, standardization played on important role in ensuring high accuracy and reliability of parts and system compatibility in mass production. Over the past 100 years, standardization has expanded to include products and manufacturing activities. Standards became important in enabling businesses to exploit their business innovations through IPR. Now a days, standardization activities are focused on the products. These are done based on the IPR owned by firms to enhance their commercialization. In these situations, standardizations efforts are started after the final stage of product development or after the local and/or national market implementation.

The evolution of standardization educations over the last 100 years can be summarized as follows:

- Standard based on their IPR is developed after development and real use in national or local market. In this approach, emphasis is given on exploiting IPR, NS and for commercial purposes and profit.

- The development of de jure and international standardization (ISz) for international market expansion and protection, based on national standard (NS).
- "How-to" approach that utilizes developed standards to facilitate and certificate on their job and developed IS education by companies or local standard bodies.
- The establishment of numerous standard bodies for every specific standard items and areas.

Nevertheless, standard setting is not without setbacks. Some standards led to adverse consequences. Some examples are:

- Tragedy of anti-commons, and/ or exclusive use of their own IPR
- GALAPAGOS syndrome: this is a situation where a local standard may lead to success in local market but fails to make a product successful overseas. As a result, the product fails to reach economies of scale.
- Not Invented Here Syndrome: some standard sets are not accepted by firms and as a result there is little buy-in and adoption. These firms give a cold response to these standards because they are not seen as reflecting the interest of others.

Recent developments show that the rapid change of technological progress is making the life of standards short.This makes the cost of developing standards and the related technologies more costly. As such, it is necessary to ensure that these standards and the related technologies are accepted and diffused quickly and extensively in markets. In addition, the WTO/TBT agreement and FRAND condition enable the fast diffusion of the IPR and enables the recovery of the cost associated with the development of the IPRs and their standards. To be more effective in doing this, global standardization should be proactive and begin much earlier i.e. during their research and development stage. This is especially important, in information and communication technology (ICT) area. This is fast becoming the trend because of the concept of software based upper compatibility and universal and LSI based digital hardware platform, those are developed by the academic and university members, such as Institute of Electric and Electrical Engineers (IEEE) and Internet Engineering Task Force (IETF) etc. De-facto, or forum standards are also becoming more important.Their standards activities are having an impact on how standard setting is done in global standardization body, such as ITU now [1]–[5]. Such a trend of from "Standardization after development" to "Standardization before development"created the paradigm shift as shown in Figure 1. This trend is also shown inshift from"international standardization

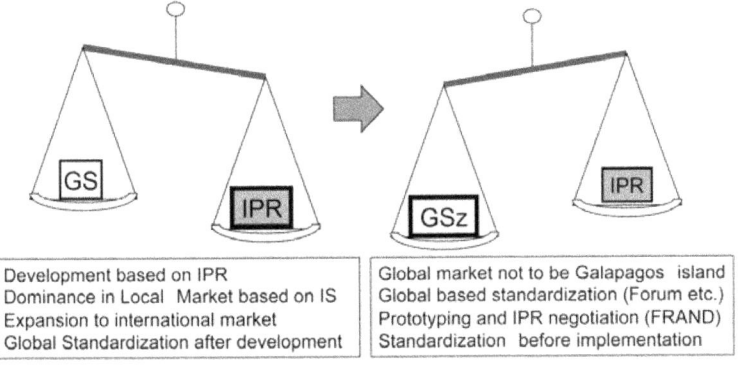

(a) Standardization of past 100 years  (b) Recent standardization after paradigm shift

Figure 1  Paradigm shift in global standardization activity.

(ISz) after setting up national standards" to "global standardization (GSz) without becoming national standards".

This shift leads to the focus away from merely the exploitation of IPR to the GS. Because of this, a discourse on standardization education among academicians and universities is needed and necessary.

"Why Education on GSz is required for academia and university" are discussed [4, 5], and setup Univ-EoGSz [11]–[12]. These trends and objectives are summarized as follows.

- GSz are on going with research and before the start of the development stage.
- Standardization activities should be run concurrently with there search stage.
- Recent standardizations are proactive. (What is required, and Why? Not How-to).
- Research activity, i.e. standardization, is essential in creating leading edge global technologies.
- Role of academians and universities in the research stage is inevitable for global standardization.
- Standardization need to begin with discussion and harmonization among community/countries first and not with companies first.
- Forum/De-facto standardization can emerged first and then move to become de jure .
- EoGSz among academia and universities, where the leading edge technology is developing, is key issue after paradigm shift.

- Role of academic members, such as IEEE etc., are important for the state of user's view
- IEICE Japan also aware about GSz activity and EoGSz.

## 2.2 Target of EoGSz and University Relevance

Target of the EoGSz will be classified by the 5 categories, as is shown in Figure 2. This section describes why university fits for these activities.

(1) Core & generic GSz Program for University Students
Education program on global standardization for academicians is better to be generic in its content, because they are not involved in specific standard items and bodies. Academic researchers and university students will start from basic and generic matter and bird watch of standardization bodies is required for them. This core and generic standardization education can

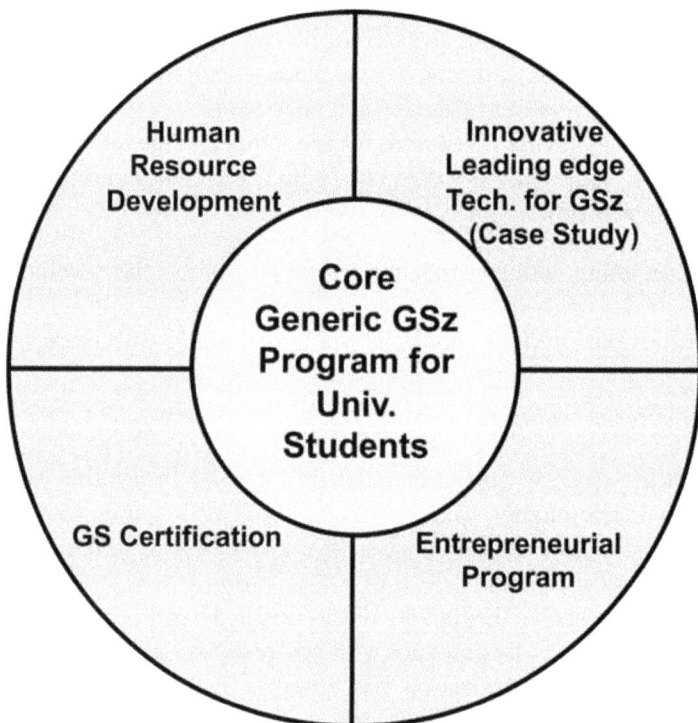

Figure 2 Objectives of education on global standardization for academic and university (Univ-EoGSz)

be incorporate in courses and programs on management of technology. In addition, basic skills required for standardization, such as communication, negotiation, IPR and legal matters, are also responsibility of business related division in university. In this case, project based learning (PBL) can be an effective method for teaching such knowledge.

(2) Innovative Leading Edge Technology (Case Study)

Academians and research staff of universities involve in research on leading edge technologies should includes standardization concerns into considerations at the research stage. Leading edge technologies such as , ex.iPS cell that was awarded Nobel Prize last year, should incorporate the standardization process to balance between social happiness and industry business. Academic members involved with this leading edge research can take the initiative to address standardization issues at the R&D stage. EoGSz is one of the powerful solutions. Knowledge of case studies of established standard and their standardization process are valuable for academic staff and students to start new standardization.

(3) Entrepreneurial Program

In various universities, entrepreneurial development programs are offered for participants for industry and students. In these programs, addressing GS and IPR issues are necessary to help participants understand how to ensure success in their entrepreneurial business. In the case of new business start-ups and the introduction of new technologies, GSz activities should be a part of the strategic thinking in planning for expanding into the global arena.This should be a part of the entrepreneurial development programs that is offered to students, businessmen, and the public in general, through the involvement and coordination between universities, industry and society.

(4) Human Resource Development for Industry and Society

In order to support the new paradigm in GSz, it is important that university students are given the necessary knowledge and know how related to new GSz. Only with this understanding will they be able to contribute to the process of incorporating standardization concerns at the research stage. This is especially important in the research activities done in universities and other public research institutions. This know how will enable to contribute towards global standards development.

(5) Global standardization Certification

In addition to the previous point, the paradigm shift in GSz requires that research organizations including universities have the necessary equipment for their R&D activities and that these equipment are certified according to global standards. Academicians and universities can also play the role of certifying labs, products and equipment based on global standards. They could be used for measuring the newly developed standard value and certification in tern. University and academia also has neutral position and regal mind to perform these certifications.

## 3 Global Standardization Education Program in Osaka Univ

The design of the Global Standardization (GSz) education program at Osaka University is based on the university's system that allows students to enroll in Major and Minor studies.

Section 3.1 describes Osaka University's standardization education program and how it is offered in relation to the Major and Minor studies. Then, design of the standardization education program is described.

### 3.1 Osaka University Graduate School Education System — Major and Minor

Osaka University consists of 17 Graduate Schools, which cover all areas of disciplines, and is based on 3 ideas on education in addition to higher scholastic achievement and specialized knowledge mastering. These 3 ideas are:

[Goals of Major studies]

(1) Comprehensive Understanding
The ability to make sound social judgments and taking a broad perspective.
(2) Synthetic Imagination
The ability to create a network that ties together people from different fields and social standings
(3) Transcultural Communicability
The ability to communicate with and understand people from various-backgrounds and cultures

To embody this, Osaka University is offering Major and Minor education system.

a. Gain a high level of specialized knowledge and develop ability to under take research

b. Gain understanding of scientific methods and the ability to utilize science

[Goals of Minor studies]
1. Credits requirement for major study completion
2. Credits requirement for minor study completion
    a. Cultivate a broad perspective of issues that integrates an interdisciplinary point of view
    b. Develop ability to understand societies and the world
    c. Develop an appreciation sense of events and opinions

Every graduate student takes their required master or doctor course of at the graduate schools where they are enrolled. Graduating requirements for master courses are 30 credits or more.

Optionally, every graduate student can take one or more of the interdisciplinary education programs.

Osaka University has been offering many interdisciplinary education programs for the minor studies since 2004. Forty six interdisciplinary education programs were offered in 2012 academic year. Each interdisciplinary education program has a subject matter and consists of multiple courses offered in different graduate schools. They are categorized into 2 education programs as in the following.

- Graduate Program for Advanced Interdisciplinary Studies

    Requirements for completion are 8 credits or more.

- Graduate Minor Program

    Requirements for completion are 14 credits or more.

The Major and Minor education system is shown in Figure 3.

## 3.2 Design of Global Standardization Education Program

It is important to understand the human resource requirements for a Global Standardization. It is not enough that instructors have the knowledge on standard setting and intellectual property. Instructors must also understand how standards setting is linked to business issues. They must also develop the abilities related to negotiation and communication with people from different background and cultures.

**(1) Designing global standardization education program for graduate students**

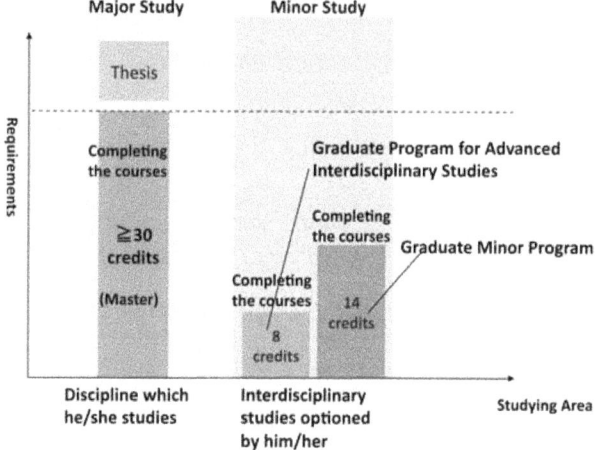

Figure 3 Major and minor education system and their credits requirements.

a. Program Target Students:
   Graduate students (reason: enough time for learning subjects and high ability of understanding for study and development)
b. Program Objectives:
   Program to offer "knowledge concerning standardization" plus "knowledge concerning negotiation and management"
c. Program Design
   The 1st step: To decide "the body of knowledge concerning global standardization" that the program would offer.

The 2nd step: To select several courses which are suitable for offering "the body of knowledge concerning global standardization.

The body of knowledge in the 1st step consists of "knowledge concerning global business, standardization, and management", "knowledge concerning standardization in information and communication fields", "knowledge concerning standardization in the manufacture field", "knowledge concerning intellectual property", "knowledge concerning knowledge value society" and "knowledge concerning negotiation".

To decide the courses in the 2nd step is done based on the body of the knowledge in the 1st step, the syllabus of the university of their own and other universities.

Figure 4. shows the courses included the Global Standardization program.

## (2) Implementation of the global standardization education program

Since 2011, the program has been offered to graduate students. Many students take courses in the program. In 2011, 212 graduate students took these courses and in 2012 another 187 graduate students took these courses. These numbers are quite satisfactory.

## (3) Evaluation of the effects of the education of the standardization program

To evaluate the outcomes of the education of the standardization program, questionnaires were sent to the students of standardization program. This consists of two questionnaires:

a. Questionnaires
   Questionnaire A: The knowledge students acquired by learning the subjects
   Questionnaire B: The likelihood they will be able to utilize the knowledge in their own carriers in the future

b. Response instruction

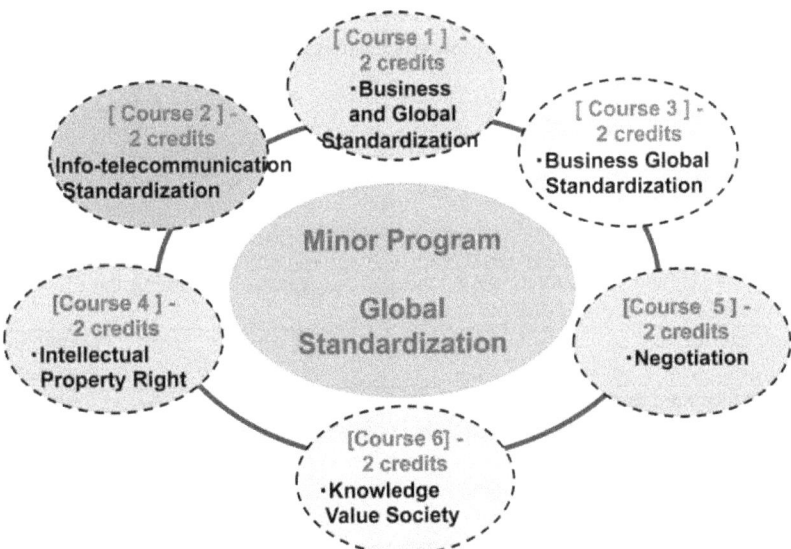

Figure 4  Courses constitution of the global standardization education program.

Students were asked to choose from several responses for both questionnaire A and B (multiple answers available)

Results of the questionnaires and answers are shown in Table 1 and can be summarized in the following.

(1) Acquired knowledge by learning the subjects; "meaning of global standardization", "items as objects of global standardization", "process to global standardization", "relationship between global standardization and corporative management strategies", "approach to global standardization", "quality of individuals necessary to approaching global standardization", "situation of approaching global standardization in Japan", "items which need new global standardization", "new research issues" and "others" are chosen as answers.

Table 1 Questionnaire results

Questionnaire A: Which type of knowledge did you gain by learning the subjects? (Multiple answers)

| No. | Answer alternative | Quantity of response |
|---|---|---|
| 1 | Meaning of global standardization | 22 |
| 2 | Items as objects of global standardization | 9 |
| 3 | Process to global standardization | 15 |
| 4 | Relationship between global standardization and corporate management strategies | 12 |
| 5 | Approach to global standardization | 13 |
| 6 | Quality of individual necessary to approaching global standardization | 8 |
| 7 | Situation of approaching global standardization in Japan | 16 |
| 8 | Items which need new global standardization | 1 |
| 9 | New research issues | 1 |
| 10 | Others | 0 |

Questionnaire B: How do you utilize the knowledge you gained by learning the subjects to your future carriers? (Multiple answers)

| No. | Answer alternative | Quantity of response |
|---|---|---|
| 1 | Planning for future research themes | 0 |
| 2 | Working for global standardization activities in the working place | 11 |
| 3 | Planning and development for products and services in the working place | 13 |
| 4 | Planning for management strategies in the working place | 7 |
| 5 | Employee training and development of human resources in the working place | 11 |
| 6 | Making use of one of your wide range of knowledge | 6 |
| 7 | Others | 0 |

(2) How to utilize the knowledge to their own carriers in the future; "planning for future research themes", "working for global standardization activities in the working place", "planning and development for products and services in the working place", "planning for management strategies in the working place", employee training and development of human resources in the working place", "making use of one of your wide range of knowledge" are chosen as answers.

c. The effect of the program
Educational contents are as well understood and the effect can be said increasing.

## 4 Education on Standardization at UTM.

### 4.1 Challenges in Introduction of Standards and Education on Standards

In the case of Malaysia, efforts to introduce standards education in universities by the Department of Standards Malaysia (DSM) has received mixed reactions from universities. Some universities, especially newer ones, are more receptive to the idea and sees standards education as a source of differentiation. Some of the universities that are more receptive are also those who have a history of offering programs in quality and standards. Those universities that are more reluctant to adopt standards educations feel that the current curriculum in their respective universities already have many course work requirements and this leaves little room to include standards education.

In addition, standards education has not inspired interest because it is largely seen to be about standards and rules, detached from business context of organizations. For many academicians, the various standards introduced by the authorities for tertiary education in Malaysia has created considerable criticism about the amount of paper work they require and the petty rules they involve. As a result, many Malaysian academicians have a less than positive view of standards and view them as a burden.

### 4.2 The Business Imperative of Standards Implementation

Choi and Behling point out that the effectiveness of quality programs is affected by the management orientation of organizations [9]. Whereas some organizations adopt quality programs to enhance their competitiveness, some do so in a minimal manner, perhaps even out of grudging compliance with customer requirements. The latter tend to be less effective in their quality programs.

Kanapathy and Jabnoun's examination of the impact of ISO certification among manufacturing firms in Malaysia did not find a positive relationship between ISO certification and performance [10]. It does appear that merely adopting standards is perhaps a necessary but not sufficient condition for many companies. Standard certification such as the ISO is necessary to enter certain markets but by itself does not ensure success in the market.

Othman and Abdul Ghani's study of the supply chain management practice of Japanese companies show that companies like Toyota do not just assess their vendors' standards compliance and certification [8]. More important to Toyota is their quality management system, including the involvement of the vendors' CEO in quality management. This shows that quality output and performance requires more than just standards compliance. Companies need to also understand how standards compliance is linked to the management system.

It is therefore necessary to rethink how standards are diffused and how the subject is taught in universities. As of now, standards education in Universiti Teknologi Malaysia is largely through a course on ISO. This will have to evolve to reflect the current understanding on competitive capabilities. Students need to not only know about standards but also how it relates to competitiveness. Only when this is done will standard education be of more value to students and their potential employers.

## 4.3 Making the Education on Standards More Relevant to Potential Adopters

In order to gain legitimacy and acceptance among companies, standards must be presented as a part of the equation in enhancing competitiveness. Its value must come from more than just a set of rules to adhere to. Businesses want to know how the adoption of standard can help with developing their capabilities. It has to ultimately be linked to the creation of better value and profitability. Standards adoption must be seen as a business issue and not merely about rule enforcement.

## 4.4 Linking Education on Standards Adoption with Business Strategy. — MJIIT-Osaka Model.

A joint discussion between faculty members from the Malaysia-Japan International Institute of Technology at Universiti Teknologi Malaysia and

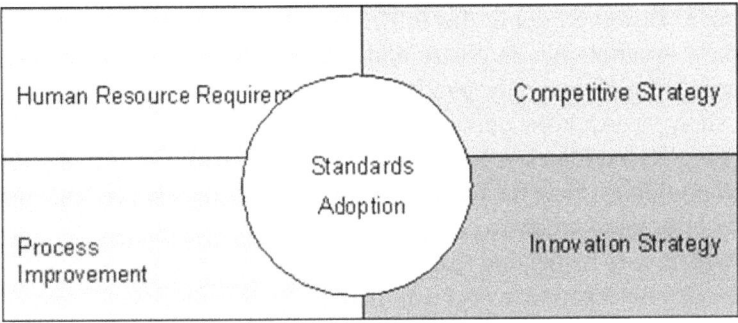

Figure 5  MJIIT-Osaka Standards Adoption Model.

Osaka University's Centre for Interdisciplinary Research and Education led to the formulation of model that reflects the need to integrate standards adoption with business strategy. Figure 5 depicts the model formulated from the discussion.

This model argues that standards adoption can only make a contribution to competitiveness when it fits the company's competitive strategy. The company's competitive strategy, in turn, defines the innovation strategy the firm should adopt. These two decisions then define the processes that need to be in place in the company and how its human resources should be developed.

For example, a Small and Medium-sized Enterprise (SME) in a developing country producing automotive parts may have little or no propriety technology. This leaves the company with no option except to compete as a low cost producer [6]. However, to be selected as a supplier, the customer would usually have a requirement that the company comply with certain standards such as the ISO and TS standards. Attaining certification is a necessary condition but it is not enough for the SME's long-term competitiveness. In addition to low cost, they need to also provide consistent quality and deliver on time [7]. To do this, this SME will need to focus on innovation that increases its efficiency. Typically, this would involve process improvement and may include adopting lean production techniques. To support all this, the SME has to ensure that its human resource has the skills needed to produce at the cost and quality desired by the customer and deliver them on time. At the same time, as a low cost producer, the SME will not invest too much in training beyond what is needed to support its production requirements.

Another example are companies producing smartphones. Change and innovation in this industry is fast and research and development need to move at a fast pace. However, doing the full range of research activities needed to create new smartphones is costly and time consuming. As such, many smart phone companies simply purchase technologies that are the de facto standards such the Android operating system and focus on differentiating by developing technologies around the operating system or in integrating acquired technologies. Here the focus of innovation is producing newer smartphones that offer more utility to customers. The process developed by such a company will have to be flexible to enable the incorporation of new technologies and close coordination with customers to understand what customers want. The human resources in these companies are expected to be creative and search for new ideas that will enable the companies to leap ahead of its competitors.

### 4.5 Scope and Framework of Standardization Education Program in MJIIT

Figure 6 shows the basic framework of the standardization education program in MJIIT. The Univ-EoGSz program will be set up as a Sub Program in the Department of Management of Technology (MOT), MJIIT for the Post Graduate students in MOT and also for PG students in the Departments of Electronic Systems Engineering (ESE), Mechanical Precision Engineering (MPE), and Environment and Green Technology (EGT). Special courses on

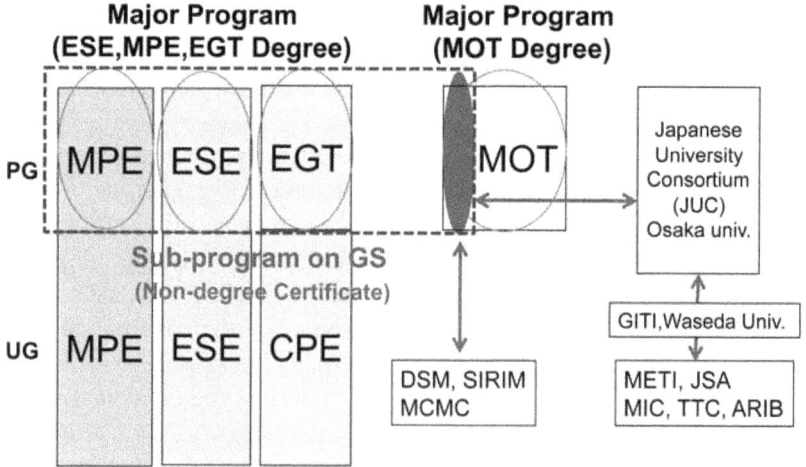

Figure 6 GS Sub-program frame structure.

Global Standardization Education Program 75

Table 2 EoGSz Sub-program frame structure and collaboration among universities.

| Course Subjects | MJIIT MOT Division, UTM Course Name | Osaka Univ. Course Name | Waseda Univ. Course Name |
|---|---|---|---|
| Generics for Global Standardization | | Global Business and Standardization | |
| Intellectual Property Right and Exercises | Malaysian Business and Intellectual Property Law | Intellectual Property Right and Exercises | |
| Devate and Negotiations | Business Networking and Managing Alliances | Project Exercise (Negotation) | |
| ISO/IEC and ITU related Global Standard Matters | | | Global Standardization for ICT |
| | | | Global Standardization of Industry Acitivity |
| Society and Marketting Related Global Standard Matters | Decision Making and Creative Problem Solving | Knowledge and Value Shared Society | |
| | Managing the Innovation Process | | |

Global Standardization will be offered in MOT in collaboration with Osaka Univ.

Osaka Univ. and MJIIT will award non-degree certificates to students who fulfill the requirements for this program (number of credit, subjects, credit transfer, etc.). The program is also supported by the Japan University Consortium (JUC) that includes 25 Japanese Universities. Some topics on GSz, related to ITU and ISO/IEC are offered by GITI, Waseda University, in collaboration with METI, JSA, MIC, TTC, ARIB in Japan.

Proposed course outline for EoGSz Sub-program is shown in Table 2. This sub-program includes generic components as well as know-how related to standardization. This is because the technology area is spreading over various divisions, as same as the situation in Osaka University's Minor Programs. The required course of EoGSz sub program is offered in a workshop format using Problem Based Learning methodology. Lectures are delivered by Osaka Univ instructors. UTM's MJIIT MOT department offers a number of courses that are similar offered in Osaka Univ., such as IPR, Negotiations, etc. These courses are included the EoGSz sub-program as elective courses. Other GSz related courses, such as decision making matter and innovation process related matter are included in this EoGSz sub-program as elective courses, as well. Those are necessary and generic skills for PBL and global standardization.

### 4.6 Relation of Malaysian Standardization Bodies and MJIIT Program

The relation ship of Malaysia's standardization bodies, such as DSM, SIRIM, and MCMC are illustrated in Figure 7.

One of the institutions designated as a standards setting organization by The Department of Standards Malaysia (DSM), Ministry of Science and Technology Innovation (MOSTI) is the Standards and Industrial Research Institute of Malaysia Berhad (SIRIM). SIRIM is a corporate organisation owned wholly by the Malaysian Government, through the Minister of Finance Incorporated. It has been entrusted by the Malaysian Government to be the national organisation for standards and quality, and has been a participating member in over 80 ISO/IEC Technical Committees and Subcommittees. SIRIM set up Industrial Standards Committees (ISC) related to technical areas, and under ISC, various Technical Committee (TC) and Working Groups (WG) exist.

In Malaysia, standards related to telecommunication and involvement in ITU matters is under the jurisdiction of the Malaysian Communication and Multimedia Commission (MCMC). The MCMC was established

Figure 7  Various Standardization Bodies relating with Malaysia and Japan.

out of the realization and explosion of a new convergent communications and multimedia industry in Malaysia in the mid 1990s, a new paradigm requiring new approaches in media policies and regulation is necessary. MCMC was established under the Communications and Multimedia Act 1998, which set out a new regulatory licensing framework for the industry and the creation of an agency to oversea. The Commission set forth 10 national policy objectives, which is the regulatory framework, which includes economic regulation, technical regulation, consumer protection and social regulation.

For the regional standardization body in Southeast Asia, the ASEAN and Plus Standards Scheme treats standardization matters. In addition, Asia-Pacific Telecommunity (APT), which was founded on the joint initiatives of the United Nations Economic and Social Commission for Asia and the Pacific (UNESCAP) and the International Telecommunication Union (ITU), serves as the focal organization for ICT in this region. The APT has 38 member countries, with 4 associate members and 130 affiliate members. Through its various programmes and activities, APT has made a significant contribution to the development and growth in the ICT sectors. The APT Wireless Group (AWG), formerly known as APT Wireless Forum (AWF) is covering various aspects of emerging wireless systems, and Asia-Pacific Telecommunity Standardization Program (ASTAP) act as to establish regional cooperation on standardization and to contribute to global standardization activities, etc.

For university and academic members in Malaysia, we are now intending to include case studies of topical, typical and leading edge technology matters, such as recycling and food rerated standards, etc. Such matters are mutually discussed with DSN now. First workshop was held in this April, and discussed the MJIIT-Osaka Standards Adoption Model shown in Figure 5, and started to setup the education on global standard and standardizations into universities in Malaysia.

Another symposium relating with recycling and their standardization will be held in this May at Osaka University with the collaboration between MJIIT and Osaka Univ.

## 5 Conclusion

This article describes the paradigm shift in global standardization, and the shift from "standardization after development" to "standardization before development". The WTO/TBT agreement and FRAND condition enable the fast diffusion of the IPR and enables the recovery of the cost associated with the development of the IPRs. To be more effective in doing this, global standardization should be proactive and done much earlier i.e. during their research and development stage.Then the GSz activity should be done in parallel or in advance during the research and development activities, which is mainly done by the academicians and university researchers. The design of the Global Standardization program is done to fulfil these requirements. This is the approach taken by the Centre for Interdisciplinary Research and Education at Osaka University. Likewise, this is also the approach taken by the MOT Department at MJIIT, UTM.

## Acknowledgement

Authors would like to express our gratitude to Prof. Ir Megat Johari Megat Mohd Noor, Dean, MJITT and member of Global Standardization Action Group, and also experess our gratitude to all the action group members for their valuable suggestions and discussions.

## References

[1] APEC Committee on Trade and Investment, "Teaching Standardization in Universities: Lessons Learned from Trial Program", APEC SCSC Education Guideline 4 - Case Book,October (2011)

[2] IEEE, "Standards Education", http://www.ieee.org/education_careers/education/standards/index.html
[3] ISO, "Education about standards", http://www.iso.org/iso/home/standards/standards-in-education.htm
[4] Komaki, S., Kobayashi, Y., "Standardization Activities of the IEICE Japan" (in Japanese), Trans on IEICE Japan, Vol.J89-B, No.2, pp55-67, (2006).
[5] ITU-T, "ITU-GISFI-DS-CTIF Standards Education Workshop", Aalborg, (2012)
[6] Choi, T. Y and Behling, O. C., "Top managers and TQM success: one more look after all these years", *Academy of Management Executive,* 11:1, pp. 37-47, (1997).
[7] Kanapathy, K. & Jabnoun, N., "Are ISO 9000 Programmes Paying off for Malaysian Manufacturing Companies?", *Malaysian Management Review,* December, pp. 40-46, (1998).
[8] Othman, R. and Abdul Ghani, R., "Supply chain management and suppliers' HRM practice". *Supply Chain Management: An International Journal,* 13:4, pp. 259-262, (2008).
[9] Tatsumoto, H., Ogawa, K. and Shintaku , J., "Standardization, international division of labour and platform business." *MMRC Discussion Paper Series,* pp.1-17, (2010).
[10] Stone, K. B. , "Four decades of lean: a systematic literature review". *International Journal of Lean Six Sigma,* 3:2, pp.112-132, (2012).
[11] Komaki, S., "Global Standardization Program of MJIIT,UTM and Osaka Univ.", ITU-GISFI-DS-CTIF Standards Education Workshop, October (2012)
[12] Nakanishi, H.,"Implementation of Education Program on Global standardization for University Students", Joint ITU-IEICE-CTIF-GISFI Workshop on Education about Standardization, April (2013)

## Biography

**Hiroshi Nakanishi** Professor, Osaka University. He was born in 1947.He graduated from graduate school of engineering of Osaka University in 1973 and got master's degree from Osaka University in 1973. He joined ECL of NTT (Electrical Communication Laboratory of Nippon Telephone and Telegraph public corporation) as a researcher in 1973. His major is electronics and information science. He had been researching and developing Magnetic and Optical storage devices, storage systems and network filing systems. Since 2006, he is working for Osaka University as a professor of Centre for Interdisciplinary Research and Education,and is researching designs of interdisciplinary education program through analysis of social needs and is teaching a program of Global Standardization.

He is a member of The Japan Society of Information and Communication Research and a member of The Institute of Electronics, Information and Communication Engineers of Japan, also a member of The Japan Society for Educational Technology.

**Rozhan Othman** Professor MJIIT, UTM. Rozhan Othman is a professor, Department Head of Technology Management, Malaysia-Japan International Institute of Technology, Universiti Teknologi Malaysia, and Chair of Standardization Action Group of MJIIT. He earned his BBA and MBA from Ohio University and his PhD from University College Dublin. His areas of research interest are HRM, Leadership and Strategic Management.

**KunioIgusa**, Professor, MJIIT, UTM. He was born in 1942. He graduated Tokyo Metropolitan University, Faculty of Economics in 1967. He joined the government research institute named IDE (Institute of Developing Economies) under the METI, Japan. He engaged research work for the government on the Asian economic policy issues and its business affairs for over 35 years. During this period, he had positioned of the Visiting researcher at Gajah Mada Univ.(1974), Research Fellow at the Indonesian Chamber of Commerce (1990), and Visiting Professor of Graduate School of Management at Macquarie Univ. (2000).

In 2001 he has appointed as the Professor of Ritsumeikan Asia Pacific Univ. (APU), being in charge of Asia Economic Management Course, and also served the Dean (2003) of the Faculty of Management of Ritumeikan APU. He has published many books, papers, articles on the Asian economicaffairs and businesses. He is invited to a professor of Management of Technology of MJIIT in 2012.

**Shozo Komaki** Professor, MJIIT, UTM. He was born in 1947. He received BS, MS and PhD degrees from Osaka University in 1970, 1972 and 1983, respectively. He joined to NTT Electrical Communication Labs.in 1972, where he was engaged in R&D on digital radio systems. From 1990, he moved to Osaka University and engaging in the research on Radio on Fiber Networks, Wireless service over IP networks, Software Definable Radio Networks and Radio Agents. He is currently a Professor of the Department of Electrical System Engineering, Malaysia International Institute of Technology, Universiti Teknologi Malaysia.

**List of Abbreviations**

| | |
|---|---|
| APT | Asia-Pacific Telecommunity |
| ASEAN | Association of South East Asian Nations and plus |
| ASTAP | Asia-Pacific Telecommunity Standardization Program |
| ARIB | Association of Radio Industry and Business |
| AWG | APT Wireless Group |
| CPE | Department of Chemical Process Engineering |
| DSM | Department of Standards Malaysia, MOSTI |
| EGT | Division of Environment and Green Technology |
| EoGS | Education on Global Standard |
| EoGSz | Education on Global Standardization |
| ESE | Department of Electronic and System Engineering |
| FRAND | Fair, Reasonable and Non Discriminatory Licensing |
| GITI | Global Information and Telecommunication Institute, Waseda University |
| GS | Global Standard |
| GSz | Global Standardization |
| ICT | Information and Communication Technology |
| IEC | International Electrotechnical Commission |
| IEEE | Institute of Electric and Electrical Engineers |
| IETF | Internet Engineering Task Force |
| IPR | Intellectual Property Rights |
| IS | International Standard |
| ISz | International Standardization |
| ISC | Industrial Standards Committees |
| ISO | International Organization for Standardization |
| ITU | International Telecommunication Union |
| JSA | Japan Standard Association |
| JUC | Japan University Consortium |
| MCMC | Malaysian Communications and Multimedia Commission |
| METI | Ministry of Economy, Trade and Industry |

**List of Abbreviations**

| | |
|---|---|
| MIC | Ministry of Internal affairs and Communications |
| MJIIT | Malaysia-Japan International Institute of Technology |
| MOSTI | Ministry of Science and Technology Innovation |
| MOT | Department of Management of Technology |
| MPE | Department of Mechanical Precision Engineering |
| RAND | Reasonable and Non Discriminatory Licensing |
| SIRIM | Standards and Industrial Research Institute of Malaysia Berhad |
| SKMM | Suruhanjaya Komunikasi Dan Multimedia Malaysia |
| SME | Small and Medium-sized Enterprise |
| TBT | Technical Barriers of Trade |
| TC | Technical Committee |
| TTC | Telecommunication Technology Commission |
| UTM | Universiti Teknologi Malaysia |
| WG | Working Group |
| WTO | World Trade Organization |

# IEEE 802.11ah: A Long Range 802.11 WLAN at Sub 1 GHz

*Weiping Sun, Munhwan Choi and Sunghyun Choi*

*Department of ECE and INMC, Seoul National University, Seoul, Korea;*
*email:{weiping, mhchoi}@mwnl.snu.ac.kr, schoi@snu.ac.kr*

Received 10 April 2013; Accepted 14 May 2013

## Abstract

IEEE 802.11ah is an emerging Wireless LAN (WLAN) standard that defines a WLAN system operating at sub 1 GHz license-exempt bands. Thanks to the favorable propagation characteristics of the low frequency spectra, 802.11ah can provide much improved transmission range compared with the conventional 802.11 WLANs operating at 2.4 GHz and 5 GHz bands. 802.11ah can be used for various purposes including large scale sensor networks, extended range hotspot, and outdoor Wi-Fi for cellular traffic offloading, whereas the available bandwidth is relatively narrow.

In this paper, we give a technical overview of 802.11ah Physical (PHY) layer and Medium Access Control (MAC) layer. For the 802.11ah PHY, which is designed based on the down-clocked operation of IEEE 802.11ac's PHY layer, we describe its channelization and transmission modes. Besides, 802.11ah MAC layer has adopted some enhancements to fulfill the expected system requirements. These enhancements include the improvement of power saving features, support of large number of stations, efficient medium access mechanisms and throughput enhancements by greater compactness of various frame formats. Through the numerical analysis, we evaluate the transmission range for indoor and outdoor environments and the theoretical throughput with newly defined channel access mechanisms.

**Keywords:** Sub 1 GHz, IEEE 802.11ah, smart grid, long range WLAN.

## 1 Introduction

IEEE 802.11 Wireless Local Area Network (WLAN), which is currently operating at 2.4 GHz and 5 GHz bands, has been one of the most popular wireless technologies in indoor environments thanks to its high data rates, easy deployment and low cost. However, the high frequency bands have put limits on the transmission ranges of the 802.11 WLAN system, and hence, the system has been treated with indifference in outdoor environments. Furthermore, the excessive utilization and irregular deployments have resulted in the saturation of the 2.4 GHz and 5 GHz spectra. The situation will worsen if more newly designed systems act as competitors.

Due to the drawbacks encountered by the current IEEE 802.11 WLANs and the increasing demands for ubiquitous wireless access, IEEE 802.11 working group has triggered a new project, called *IEEE 802.11ah*, that attempts to enact an 802.11 standard at sub 1 GHz license-exempt bands for cost-effective and large scale wireless networks. As described in [1], within the 802.11 working group, there are five steps in developing a new standard amendment: 1) discussion of new ideas in the Wireless Next Generation Standing Committee, 2) development of the purpose and scope for the amendment in a study group, 3) drafting an amendment in a task group, 4) approval of the draft by the working group, and 5) review by a sponsor ballot pool, and approval and ratification of the draft by the IEEE Standards Association board. The sub 1 GHz standardization activity is currently in step 3, and the corresponding task group is called *Task Group ah (TGah)*.

A critical deficiency encountered by the 802.11ah system is the scarcity of the available spectra in the sub 1 GHz *Industrial, Scientific, and Medical (ISM)* bands, such that increasing the spectral efficiency has been one of the main concerns in system design. In order to increase the system throughput, TGah has designed a new *Physical (PHY)* layer based on the *IEEE 802.11ac* [2], which is another amendment of the 802.11 family designed for high throughput WLAN. Besides, apart from the PHY layer design, there also have been some efforts made in *Medium Access Control (MAC)* layer to increase the system throughput. Moreover, when designing a new 802.11 system at sub 1 GHz bands, there is no need to maintain the backward compatibility with the other 802.11 systems, since the new system will operate at entirely different frequency bands, thus allowing the 802.11ah to define some new compact frame formats to reduce the protocol overhead without considering the backward compatibility.

On the other hand, thanks to the favorable propagation characteristics of such a low frequency spectrum, the 802.11ah system is supposed to provide much improved transmission range compared with the current 802.11 WLANs operating at 2.4 GHz and 5 GHz bands. Low-cost and large coverage properties make the 802.11ah system highly attractive for large scale sensor networks, e.g., smart grid, in which the number of involved devices in a given network could be much larger than that of the current 802.11 system. On the other hand, the target devices in the sensor networks are likely to be battery-powered devices, and hence, the power saving features become much more critical to the performance of 802.11ah system. In order to cope with such expected requirements, some enhancements have been considered in 802.11ah MAC layer design in terms of power saving and support of large number of stations.

In this paper we first introduce the current activities and status of IEEE 802.11ah project, and then elaborate the use cases of 802.11ah for helping the understanding of the intention of its various techniques. Afterwards, we will describe the features related with PHY layer in terms of *channelization* and *transmission modes*, which are based on the 10 times down-clocked version of 802.11ac PHY. MAC layer techniques, for which the agreements are already made as part of the 802.11ah draft, will be described from four aspects, namely, *support of large number of stations*, *power saving*, *channel access*, and *throughput enhancements*. We also provide the performance of 802.11ah in terms of transmission range and theoretical MAC layer throughput.

The remainder of the paper is organized as follows. Section 2 presents the current activities and status of IEEE 802.11ah project. In Section 3, the 802.11ah use cases are elaborated. Afterwards, PHY and MAC layer design issues as well as the resulting techniques are presented in Section 4 and Section 5, respectively. The performance evaluation results are presented in Section 6, and finally, the paper concludes with Section 7.

## 2 Current Standardization Activity

After *Sub 1 GHz Study Group* had completed its primary work of generating a *Project Authorization Request (PAR)* document [3], which describes the purpose and scope of the IEEE 802.11ah project, the standardization work was undertaken by TGah in November 2010. The first step in drafting an amendment was to develop a *selection procedure* [4], which would be executed and followed by TGah till the 802.11ah draft specification is complete and coherent enough for a working group letter ballot. Besides, TGah has adopted *usage models* [5], *channel models* [6], and *functional requirements* [7], based on which the evaluation of the incoming technical submissions have been conducted. A *Specification Framework Document (SFD)* [8] was created by TGah, which outlines the main functional blocks of the proposed specification. The SFD is still evolving by including the contributions of the incoming technical submissions, which are adopted based on the procedure specified in [4]. Two *Functional Block Ad Hoc Sub Groups* were created, namely *PHY Ad Hoc Sub Group* and *MAC Ad Hoc Sub Group*, being responsible for the drafting specifications of the two main functional blocks, i.e., PHY layer and MAC layer. TGah alternately holds face to face plenary and interim sessions every two months, during which chairs of each Ad Hoc Sub Groups report on their progress and content to the entire task group. These update sessions provide the opportunity for peer review to ensure the creation of a coherent specification [4].

So far, TGah has faced a number of major technical issues, such as long range transmissions, increased power saving features, support of large number of stations, and throughput enhancements, most of which have already been resolved. However, there are still some remaining technical issues which are needed to be carefully addressed. For example, there has been a proposal [9] advocating inclusion of *two-hop relaying* concept in 802.11ah, and the idea has been adopted and is currently under improvement. By utilizing the relay-based two-hop transmissions, the service range of an AP can be extended and the energy efficiency of the cell-edge stations can be enhanced. TGah is currently considering further relay-specific operations and attempting to mitigate some side effects, e.g., the increased contentions for channel access. Besides, since the coverage of a 802.11ah AP can be substantially large, the performance of a certain 802.11ah network can be severely affected by the interference generated by neighboring networks. For the aforementioned problem, an effective solution adopted by TGah is to employ beamforming techniques to divide the whole network into several sectors and use a *Time-Division Multiplexing (TDM)* approach to spread the transmissions of different sectors [10]. Standardization on the sectorized transmissions is still on-going.

The generation of IEEE 802.11ah draft 1.0 will be followed by an initial letter ballot within the 802.11 working group, which is expected to be conducted in September 2013. After all the technical issues are resolved through a number of letter ballots, an initial sponsor ballot is expected in March 2015. The standardization should be completed approximately by March 2016.

## 3 Use Cases

The characteristics of 802.11ah makes it attractive for various purposes. The general categories of the use cases include *sensors networks*, *backhaul networks for sensor and meter*, and *extended range Wi-Fi* [5]. The following discussions will deliver the descriptions of two major use cases, i.e., *sensors networks* and *backhaul networks for sensor and meter* to help the understanding of the advantages of using sub 1 GHz bands in various domains and scenarios.

In order to make the public utility greener, more and more utility companies start deploying a large number of wireless sensors and meters around their utility infrastructures. Such an electrical grid is called *smart grid*, whose functions are to monitor the real-time status of various utility consumptions and inform the company and end-users of their usage status, e.g., gas, water and power consumptions [11]. Typically, the number of devices involved in smart grid is much higher than that in traditional 802.11 WLANs, and the required transmission range of the involved devices is also much wider than that in traditional 802.11 WLANs.

In sub 1 GHz system, owing to the improved propagation feature, the coverage of one-hop transmission can be much wider, thus allowing to support more devices in a single network. As such, IEEE 802.11ah has included the application of sensors and meters as one of the major use cases [5]. Figure 1(a) shows a simple smart grid scenario. In the proposed scenario, there is an 802.11ah AP placed at outdoor area, and the stations, such as gas meter, power meter, and water meter, are deployed in indoor area. Besides, *distributed automation devices*, whose role is to increase the reliability and utilization of the existing infrastructure, are deployed at outdoor regions. In outdoor area, the coverage of the AP up to 1 km is required [3], whereas at least 100 kbps data rate is assumed in this use case [5].

Another use case is the backhaul connection between sensors and/or data collectors. In this case, IEEE 802.15.4g [12] is supposed to provide a link for lower traffic leaf sensor and IEEE 802.11ah is going to provide a wireless backhaul link to accommodate the aggregated traffic generated by the leaf sensors. Besides, the large coverage of sub 1 GHz allows a simple network design to link IEEE 802.11ah APs together, e.g., as wireless mesh networks [13]. Figure 1(b) illustrates a wireless backhaul network, composed of IEEE 802.11ah AP and gateways, which aggregate and forward the traffic from sensor devices, e.g., 802.15.4g devices, to remote control and data base.

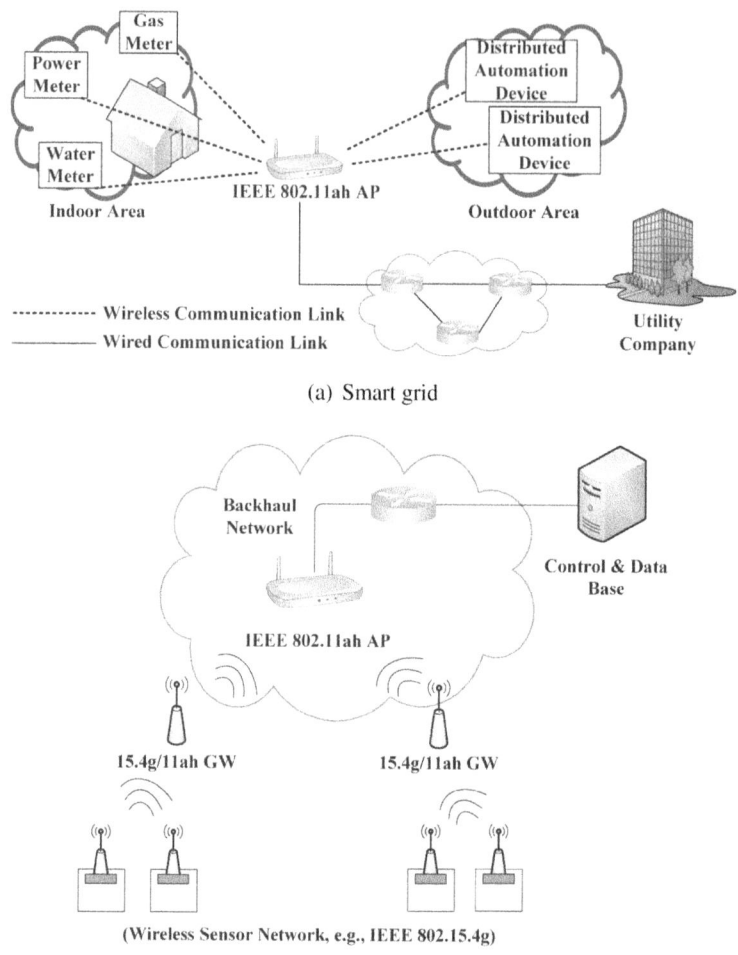

(a) Smart grid

(b) Backhaul networks for sensor and meter

Figure 1  IEEE 802.11ah use cases.

## 4 PHY Layer

Regarding the 802.11ah PHY features, which are based on the down-clocked operation of IEEE 802.11ac's PHY, we describe its channelization and various types of transmission modes. As an evolution of 802.11n, to achieve higher data rates, 802.11ac provides 80 MHz, 160 MHz and non-contiguous 160 MHz channel bandwidths in addition to the 802.11n's 20 MHz and 40 MHz channel bandwidths. Being a 10-times down-clocked version of 802.11ac, IEEE

802.11ah defines 2 MHz, 4 MHz, 8 MHz, and 16 MHz channels. Besides, 1 MHz channel is additionally defined by 802.11ah for the purpose of extended coverage. In the following, we will provide 802.11ah's PHY features in terms of channelization and transmission modes.

## 4.1 Channelization

The available sub 1 GHz ISM bands are different depending on countries, and hence, IEEE 802.11ah has defined the channelization based on the respective available wireless spectra in various countries, including the United States, South Korea, China, Europe, Japan, and Singapore. In the following, we outline the 802.11ah channelization in these countries.

As a representative example, we illustrate the channelization in the United States in Figure 2. Total 26 MHz band between 902 MHz and 928 MHz is available in the US, and accordingly, the number of available 1 MHz channels is 26. In order to achieve a higher bandwidth, 802.11ah maintains the same channel bonding method as in 802.11n and 802.11ac, i.e., several adjacent narrower channels are bonded together to yield a wider channel. As a result, 2 MHz channel is composed of two adjacent 1 MHz channels. Similarly, more wider channel bandwidths are supported through channel bonding. The widest channel supported in the US is 16 MHz channel, which is also the widest channel bandwidth supported in the 802.11ah system.

Figure 3 shows the sub 1 GHz spectra specified in the 802.11ah channelization with respect to the involved countries. The channelization for South Korea, which starts from 917.5 MHz and ends at 923.5 MHz, defines 6 MHz bandwidth. The reason for the 0.5 MHz frequency offset is to reduce the

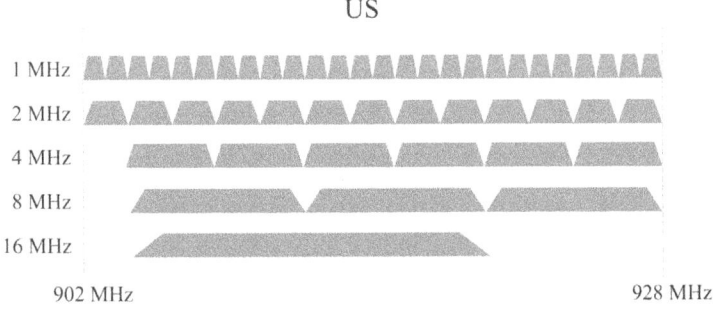

Figure 2 IEEE 802.11ah channelization for the US.

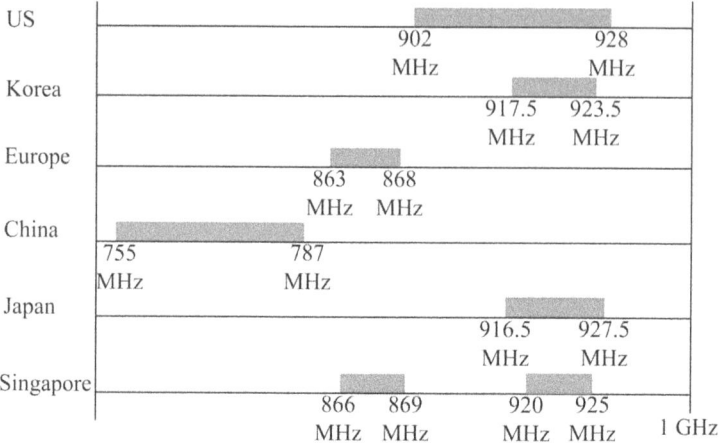

Figure 3 Sub 1 GHz spectra specified in the 802.11ah channelization.

possible mutual interference with wireless legacy systems at lower frequencies [13]. The available spectra defined in the 802.11ah channelization for Europe, China, Japan are 863–868 MHz, 755–787 MHz, 916.5–927.5 MHz, respectively, and in Singapore, the specified spectra are composed of two non-contiguous frequency bands, i.e., 866–869 MHz and 920–925 MHz bands. Besides, in Japan, there is also a 0.5 MHz offset, because Japanese spectrum regulations specified the center frequencies of the spectrum instead of start/stop frequencies [14]. These spectra are channelized based on the same rules as shown in the case of the US, i.e., split the radio bands into multiple 1 MHz channels and achieve wider channel bandwidths through channel bonding.

However, since various involved countries have different regulations and available spectra, the maximum channel bandwidths obtained by the channel bonding could be different. Accordingly, the maximum channel bandwidths supported by South Korea, Europe, China, Japan, and Singapore are set to 4 MHz, 2 MHz, 8 MHz, 1 MHz, and 4 MHz, respectively. More detailed information for these countries' channelization is available in [8].

## 4.2 Transmission Modes

In 802.11ah, 1 MHz and 2 MHz channels have been adopted as common channel bandwidths such that 802.11ah stations have to support the receptions of them. The PHY layer design can be classified into 2 categories. The first category is the transmission modes of greater than or equal to 2 MHz channel bandwidths and the other is the transmission mode of 1 MHz channel bandwidth.

For the greater than or equal to 2 MHz modes, i.e., 2 MHz, 4 MHz, 8 MHz, and 16 MHz transmissions, the PHY layer is exactly designed based on 10 times down-clocking of 802.11ac's PHY layer. That is, techniques like *Orthogonal Frequency Division Multiplexing (OFDM)* and *Multi Input Multi Output (MIMO)* have been adopted, and Downlink Multi-User MIMO (DL MU-MIMO), which is firstly introduced in the 802.11ac, is also employed by the 802.11ah system. Besides, in 802.11ah, an OFDM symbol duration is exactly ten times of that of 802.11ac, and the number of data tones in 2 MHz, 4 MHz, 8 MHz, and 16 MHz channels in 802.11ah are the same as those in 20 MHz, 40 MHz, 80 MHz, and 160 MHz channels in 802.11ac. The set of supported MCSs is also the same as that of 802.11ac.

Table 1 shows the MCSs and the corresponding data rates using 2 MHz channel with a single spatial stream. $N_{SS}$ represents the number of spatial streams, which is 1 in this case. $N_{SD}$ denotes the number of subcarriers used in data transmission. In the 2 MHz channel, 64 *Fast Fourier Transform (FFT)* is used to generate an OFDM symbol, and among the 64 subcarriers, the number of subcarriers used to transmit data is 52. $N_{DBPS}$ indicates the number of data bits per symbol, which is calculated by the number of data bits per subcarrier per symbol multiplied by the number of data subcarriers. The right-most column represent the corresponding data rates, which are calculated as the number of symbols per second multiplied by $N_{DBPS}$. A *Guard Interval (GI)* is a portion of an OFDM symbol containing redundant data, being used to prevent *Inter-Symbol Interference (ISI)* in OFDM transmission. In the case of adopting short GI of 4 $\mu s$, an OFDM symbol duration becomes 36 $\mu s$, whereas, when using normal GI of 8 $\mu s$, the OFDM symbol duration becomes 40 $\mu s$. Consequently, the data rates achieved by adopting short GI results in approximately 11 % increase in data rates compared to the case of adopting normal GI. Moreover, according to the MCS exclusion-rules of 802.11ac, MCS 9 is not adopted in 20 MHz channel with a single spatial stream. Similarly, in the case of 802.11ah 2 MHz channel with a single spatial stream, the MCS 9 is not valid either.

Those specifications result in that the data rates of 802.11ah are exactly one-tenth of 802.11ac's data rates due to the ten times extended symbol duration. For instance, when using a single spatial stream and MCS 0 in 2 MHz channel, the data rate is 0.65 Mbps in 802.11ah, which is exactly one-tenth of the data rate achieved in 802.11ac. Moreover, in order to take balance between throughput and power consumption, the maximum number of spatial streams

Table 1 802.11ah MCSs and data rates for 2 MHz channel, $N_{SS} = 1$

| MCS Index | Modulation | Code Rate | $N_{SD}$ | $N_{DBPS}$ | Data Rate (Mbps) Normal GI | Data Rate (Mbps) Short GI |
|---|---|---|---|---|---|---|
| 0 | BPSK | 1/2 | 52 | 26 | 0.65 | 0.72 |
| 1 | QPSK | 1/2 | 52 | 52 | 1.3 | 1.44 |
| 2 | QPSK | 3/4 | 52 | 78 | 1.95 | 2.17 |
| 3 | 16-QAM | 1/2 | 52 | 104 | 2.6 | 2.89 |
| 4 | 16-QAM | 3/4 | 52 | 156 | 3.9 | 4.33 |
| 5 | 64-QAM | 2/3 | 52 | 208 | 5.2 | 5.78 |
| 6 | 64-QAM | 3/4 | 52 | 234 | 5.85 | 6.5 |
| 7 | 64-QAM | 5/6 | 52 | 260 | 6.5 | 7.22 |
| 8 | 256-QAM | 3/4 | 52 | 312 | 7.8 | 8.67 |
| 9 | 256-QAM | 5/6 | | | Not valid | |

supportable in 802.11ah is up to 4, whereas in 802.11ac, a device can support up to 8 spatial streams.

For the 1 MHz transmission mode, 802.11ah maintains the same tone spacing as in the case of the former transmission modes, which is 31.25 kHz, resulting in 32 *Fast Fourier Transform (FFT)*, whereas 64 FFT is used in 2 MHz transmission. However, in 1 MHz channel, the number of data subcarriers per OFDM symbol is 24, which is less than a half of data subcarriers in 2 MHz channel. The reason is that when two adjacent narrower channels are bonded together to yield a wider channel, the number of data subcarries become more than double since the guard band between the two bonded channels can be removed.

The goal of designing 1 MHz channel is to extend the transmission range. To this end, a new MCS index, which is called *MCS 10*, is included for long range transmission in addition to the 802.11ac's MCSs. This MCS is nothing but a mode of MCS 0 with 2x repetition, by which the transmission range can be enlarged since the symbol repetition increases the reliability of the wireless transmission further.

## 5 MAC Layer

In 802.11ah MAC layer design, some features are enhanced compared with the existing 802.11 MAC, including improvements related with support of large number of stations, power saving, medium access mechanisms and throughput enhancements. In the following, these MAC layer enhancements are presented.

## 5.1 Support of Large Number of Associated Stations

In 802.11 system, an AP allocates an identifier called *Association IDentifier (AID)*, to each station during the association stage [15]. In a given network, AID is a unique ID, through which the AP can indicate its associated stations. The possible number of associated stations of an AP is up to 2,007 in legacy 802.11 standard due to the limited length of the *partial virtual bitmap* of *Traffic Indication Map (TIM) Information Element (IE)*, where each bit indicates the corresponding station's AID. The TIM IE is used to support stations' power management, and more descriptions about it will be provided in the next sub-section.

As described in Section 3, in IEEE 802.11ah system, an AP is likely to be associated with much more stations than that in legacy 802.11 networks, and hence, 802.11ah has increased the number of supportable stations to meet such expected requirements.

For increasing the number of supportable stations, a hierarchical AID structure is newly defined in 802.11ah as illustrated in Figure 4. It consists of 13 bits, and accordingly, the number of stations that it can express is up to $2^{13} - 1 (= 8,191)$. It is composed of four hierarchical levels, namely, *page*, *block*, *sub-block*, and *station's index in sub-block*. That is, each station belongs to a certain sub-block, and each sub-block belongs to a certain block. Similarly, multiple blocks form a page, which is the highest level that can contain up to 2048 stations. This hierarchical AID structure enables us to indicate more stations' AID with a given length of partial virtual bitmap. For example, when we need to indicate multiple stations, we can simply include them in a block or

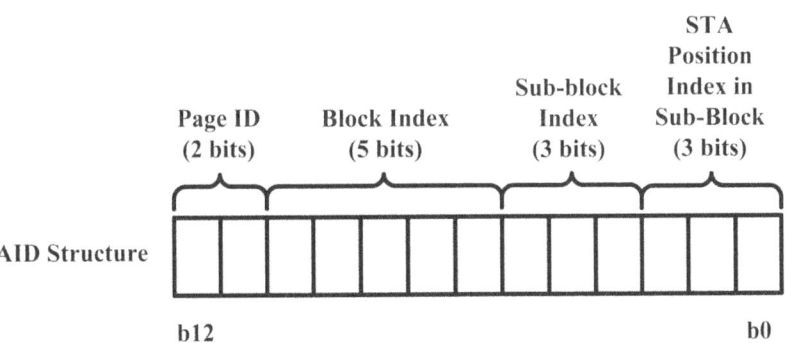

Figure 4 AID structure.

a sub-block and use the block ID or the sub-block ID to indicate them instead of including all of their AIDs.

Furthermore, there could be stations with different traffic patterns and/or different locations. If we can easily group these stations based on some specific properties, the wireless resource could be utilized more efficiently. There have been many existing research efforts on clustering algorithms. As a well-known principle, hierarchical structure can be adopted to facilitate the grouping so that the characteristic of hierarchical AID structure makes grouping of the stations much easier. Grouping can be used for various purposes, such as power saving enhancements, resource allocation, and efficient channel access, which will be detailed later.

## 5.2 Power Saving

The legacy 802.11 standard defines two power management modes. With *active* mode, a station continually turns on the radio components, i.e., *awake* state, such that it senses the incoming signal all the time and can also transmit and receive signals. On the other hand, with *power saving mode*, a station alternates between *awake* state and *doze* state, where the station in doze state turns off the radio components such that it cannot sense incoming signals at all. When there are packets destined to a station in doze state, the AP buffers the packets until the station wakes up and requests the delivery of the buffered traffic.

In 802.11 system, AP periodically transmits beacon frame, which contains TIM IE. The partial virtual bitmap field in the TIM IE conveys the information of the existence of buffered traffic destined to power saving stations. Accordingly, a power saving station needs to wake up periodically to receive a beacon, based on which it can check the existence of the buffered packets destined to itself. If it recognizes the existence of buffered traffic, the station then transmits a control frame called *Power Saving (PS)-poll frame* to the AP to request the delivery of the buffered packets. After finishing the reception of the buffered packets, the power saving station can go back to doze state.

However, several undesirable phenomena could happen, when there exist a large number of stations in a network. One is that the length of the beacon frame could become extremely long due to the excessive length of the partial virtual bitmap in TIM IE. In addition, if the amount of the buffered traffic is too heavy to be accommodated within a beacon interval, some power saving stations inevitably keep in awake state to complete the receptions of their buffered packets.

In order to solve the aforementioned problems, 802.11ah introduces a mechanism called *TIM and page segmentation*, which works as follows. Firstly, an AP splits the whole partial virtual bitmap corresponding to one page into multiple page segments, and each beacon is responsible for carrying the buffering status of only a certain page segment. Then, each power saving station wakes up at the transmission time of the beacon which carries the buffering information of the segments it belongs to. A new IE called *segment count IE* is defined to deliver segmentation information, such as the resulting number of segments after page segmentation and the boundary of each page segment.

DTIM beacon is a beacon frame that includes *Delivery TIM (DTIM) IE*, which is a special type of TIM IE used to indicate the buffering status of group addressed packets (i.e., multicast and broadcast packets). Right after a DTIM beacon transmission, the AP transmits all the pending group addressed packets. Generally, all of the power saving stations wake up to receive DTIM beacons so that they do not miss group addressed packets. In order to let power saving stations know the segmentation information beforehand, the segment count IE is contained in every DTIM beacon which is transmitted periodically over several beacon intervals.

Figure 5 illustrates the usage of the TIM and page segmentation. We assume that a page is composed of 32 stations and the page is divided into 4 page segments since one DTIM beacon interval consists of 4 beacon intervals in this example. Each beacon sequentially carries each page segment, and each beacon interval is used to accommodate traffic delivery of the corresponding page segment. Power saving stations wake up in time to receive the DTIM beacon and then through the information conveyed in segment count IE, they can recognize which beacon contains the partial virtual bitmap of the page segment that they belong to. Afterwards, power saving stations can go back to doze state and wake up again at the transmission time of the beacon frame they should refer to.

As a result, the length of each beacon frame can be shortened because the partial virtual bitmap field of the TIM IE in this case only indicates a certain segment instead of the whole page. Besides, after the reception of the DTIM beacon, power saving stations can only be awake for the beacon with their affiliated segments, and stay in doze state during the other beacon periods, through which the unnecessary energy consumption during excessive awake state can be saved.

In 802.11ah system, power saving stations are categorized into two classes. The first class is called *TIM station*, which is similar to the concept of power

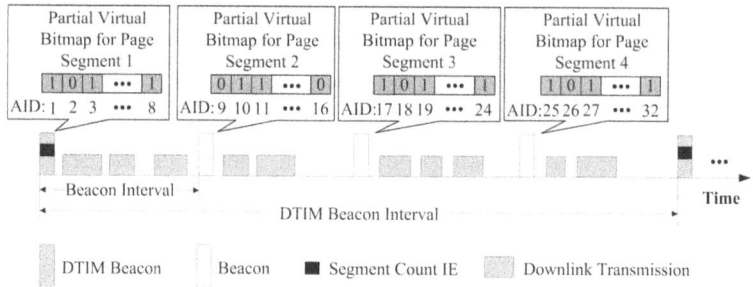

Figure 5  An example diagram of page segmentation.

saving mode in legacy 802.11 systm. That is, the packet buffering information of these stations is included in TIM IE. On the other hand, especially for low-powered sensor devices, 802.11ah defines another power saving mode, in which the buffering information is not included in TIM IE assuming that there is no need for them to periodically wake up for the beacon reception. The station operating in this mode is called *non-TIM station*, and by operating as a non-TIM station, the station can further save the energy consumption compared with the TIM stations, because it can keep in doze state over a longer period without worrying about beacon reception.

### 5.3 Channel Access

802.11ah has introduced some novel channel access mechanisms for both TIM stations and non-TIM stations.

For non-TIM stations, AP may allow them to request buffered downlink traffic or to transmit uplink traffic at anytime upon waking up. However, in such an arbitrary manner, there could be much uncontrollable traffic incurred by these non-TIM stations, which is likely to degrade the network performance. For example, if a large number of stations wake up at the same time, the contention among these stations could results in excessive channel access delays or even collisions.

To make the non-TIM stations' traffic under control, 802.11ah AP can let them wake up at a predefined time so that the wake-up time of these non-TIM stations and their channel access attempts could be temporally spread out. To exchange the wake up timing information between AP and stations, 802.11ah has defined an IE called *Target Wake Time (TWT) IE*, which is exchanged by *association request* and *association response* frames. In TWT IE, there are four fields, i.e., *request type, target wake time, minimum wake duration,* and *wake interval mantissa*, which are used to determine when and how often a station wakes up for downlink and/or uplink transmissions.

More specifically, when there are buffered packets for a non-TIM station, the AP can send to the station a newly defined control frame called *Null Data Packet (NDP) paging* frame at its target wake time, which contains the information of buffering status. If the station recognizes the existence of buffered packets after successfully receiving the NDP paging frame, it can then request the delivery of the buffered packets by transmitting a PS-poll frame. If the NDP paging frame is not transmitted by the AP at the target wake time, the station can transmit uplink frame if the channel is idle.

For TIM stations, in addition to the existing contention-based channel access mechanisms, e.g., *Enhanced Distributed Channel Access (EDCA)*, 802.11ah has focused its efforts on defining a new type of contention-free channel access mechanism, which is motivated by the increased contention level and severity of hidden terminal problems due to the increased number of stations involved in a network.

As a result, a concept named *Restricted Access Window (RAW)* has been proposed. An RAW is a time duration composed of several time slots. An AP may indicate to a TIM station a time slot during which the station is allowed to transmit or to receive packets. RAW can be used for various purposes. For example, it can be allocated to a group of TIM stations that have uplink or downlink data packets or be reserved for control frames, e.g., PS-poll.

Besides, in order to indicate the parameters related with RAW allocation, e.g., RAW start time, RAW duration, and the AIDs of the stations to which the RAW is allocated, *RAW Parameter Set (RPS) IE* is proposed. The RPS IE is optionally included in beacon frame, and each station can recognize the allocated RAW via the RPS.

Moreover, it would be more efficient to allocate time slots only to the stations which are certainly ready to transmit or receive, rather than to all of the TIM stations. In order to enable the AP to adaptively manage the RAW allocation, a new management frame named *Resource Allocation (RA) frame* has been proposed, which contains the scheduling information of each individual station, through which the station can learn the time slot during which it is allowed to conduct medium access for uplink or downlink transmission. The RA frame is transmitted at the beginning of each RAW and all the stations assigned to that RAW have to wake up to receive it. 802.11ah adds a special field called *Uplink Data Indication (UDI)* in the PS-poll frame. The UDI is used to indicate the existence of the uplink frame of a station, and a station with no buffered downlink frame, can send the PS-poll with UDI field set to 1, to request the time slot for its uplink transmission. After receiving both normal PS-poll and UDI PS-poll, the AP then can determine how to schedule the uplink and downlink transmissions efficiently, and the scheduling information is contained in the RA frame.

Figure 6 illustrates the uplink and downlink packet delivery procedures by applying the RAW concept. During the first RAW, i.e., RAW 1, AP allocates each TIM station a time slot for PS-poll transmission. We denote the PS-poll with UDI field set to 1 as *UDI* to differentiate it from a normal PS-poll frame. The station with AID 1, which does not have buffered downlink traffic, sends a UDI to request time resource for its uplink transmission, and the stations with AID 2, AID 3, and AID 5 transmit normal PS-poll frames during the allocated time slot except the station with AID 4. Such unexpected behavior of the station with AID 4 in this example, i.e., not transmitting a normal PS-poll frame, can occur due to the stations' asynchronous operation with its associated AP or the erroneous reception of a beacon frame. During RAW 2, firstly, the AP transmits an RA frame for delivering the scheduling information for stations with AID 1, AID 2, AID 3, and AID 5, and thereafter, these stations conduct uplink or downlink transmissions during the allocated time slots, respectively.

### 5.4 Throughput Enhancements

As described earlier, a major deficiency of 802.11ah is its low data rates, and to overcome it, there have been many efforts on throughput enhancements.

The first intention was to design more compact frame formats to reduce protocol overheads, and correspondingly, a more compact MAC header format has been proposed. Figure 7 compares the proposed short MAC header format

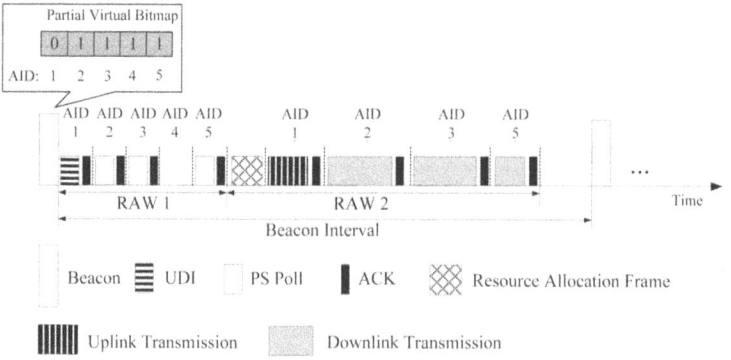

Figure 6 An example of uplink and downlink data delivery using RAW.

and the legacy 802.11 MAC header format. For the downlink, the address 1, which is destination MAC address in the legacy MAC header, is replaced with the AID of the destination station in the short MAC header format. Similarly, in the case of uplink, the address 2, which is the MAC address of source station in the legacy MAC header, is replaced with the AID of source station in the short MAC header. In addition, in the short MAC header, the sequence control field is moved before address 3 field. The address 3 field is optionally included in the short MAC header and its inclusion is indicated by the indication bit in *Frame Control (FC)* field.

Moreover, some necessary information contained in *Quality of Service (QoS)* field and *High Throughput (HT)* field is moved to *Signal (SIG)* field in the PHY header and the other unnecessary parts are removed such that there is no QoS and HT fields presented in the short MAC header. Another thing to note here is that there is no *duration/ID* field in the short MAC header so that *virtual carrier sensing* is not supported when using short MAC header. By substituting the 6 bytes MAC address with 2 bytes AID and eliminate duration/ID, QoS, and HT fields, it is possible to save at least 12 bytes overhead in both uplink and downlink. This kind of short MAC header should be used after the AID assignment procedure, and is differentiated from the legacy MAC header by setting a new *protocol version* value in the FC field.

In legacy 802.11 standard, *Acknowledgement (ACK)* frame includes MAC header and *Frame Check Sequence (FCS)* field in addition to the preamble of the packet. In order to shorten the ACK frame, 802.11ah has proposed a new ACK frame format called *Null Data Packet (NDP) ACK*, in which the

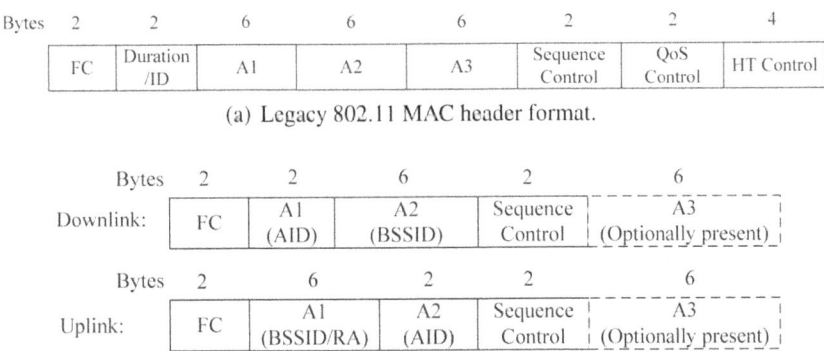

Figure 7 MAC header comparison between 802.11 legacy system and 802.11ah system.

| STF | LTF1 | SIG |
|---|---|---|

Figure 8  NDP ACK frame format.

MAC header and FCS field are eliminated such that the frame only contains PHY header field as illustrated in Figure 8. Besides, the NDP ACK frame is identified by a reserved value of MCS, which is indicated in the SIG field of the PHY header. There are also some other control frames modified to an NDP frame format to reduce the protocol overhead, e.g., *Block ACK, Clear To Send (CTS)*, and PS-poll frames, and these NDP control frames are indicated by other reserved MCS values as in the case of the NDP ACK.

Besides, 802.11ah defines a novel medium access mechanism, by which the channel access delay and ACK transmission overhead are eliminated so that the achieved throughput is increased. In legacy 802.11 standard, the feedback method of a receiver can be requested by the transmitter through *ACK indication* field. The 2 bits ACK indication field can express 4 different values, three of which had been already defined to indicate feedback methods of *normal ACK, block ACK*, and *no response*, respectively, while the last one had been reserved for future usage.

In 802.11ah, the reserved value is defined to indicate another feedback method. With this method, if the receiver has a frame destined to the transmitter, it can notify the successful reception by transmitting its data frame instead of ACK or block ACK. Similarly, if the transmitter successfully receives the receiver's data frame, it can also reply with its data frame instead of other control frames, and the gap between each transmission is restricted to *Short Inter-frame Space (SIFS)*. Such a method is called *speed frame exchange*. It speeds up the interchange of frames between AP and stations, because the ACK overhead and channel access delays are eliminated. Figure 9 illustrates an example of sequential transmissions between an AP and a station using speed frame exchange. When a transmitter, i.e., station or AP, sets ACK indication bits to request a data frame as its feedback of successful transmission instead of other feedback methods, the corresponding receiver, i.e., AP or station, replies with data frames until there is no packet to transmit. By applying this method, besides the throughput gain we can obtain by reduced overhead, the power saving stations can save more energy, because in this case, the time for awake state is reduced compared to the normal data transmissions. Moreover, the speed frame exchange scheme is more effective when there are a similar number of uplink and downlink packets, since the

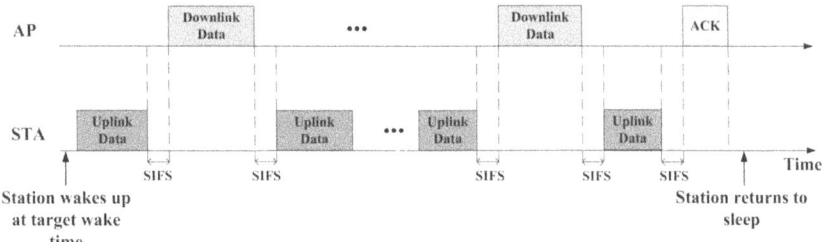

Figure 9 Speed frame exchange.

ACK transmission could not be replaced unless there are more data packets available.

## 6 Performance Evaluation

As we mentioned above, 802.11ah provides large transmission coverage, while there have been some efforts on reducing the protocol overheads to overcome its weakness of throughput. To verify these aspects, in this section, the performance of 802.11ah system is evaluated in terms of its transmission range and throughput.

### 6.1 Transmission Range

We will evaluate the transmission range by comparing the performance of 802.11ah and current 2.4 GHz and 5 GHz 802.11 systems. The transmission range is calculated by only considering path loss effect. At the receiver side, the minimum input level sensitivities specified in 802.11ah, 802.11n, and 802.11ac standard specification are used to determine the minimum received power levels required for successful decoding in 900 MHz, 2.4 GHz, and 5 GHz bands, respectively. Besides, we assume that the transmissions are conducted with a single spatial stream. As proposed in [6], we adopt *TGn channel model* and *3GPP cellular system simulation channel model* in indoor and outdoor environments, respectively. The specific calculating methods are described in [16, 17].

Figure 10 shows the transmission range in indoor and outdoor environments at various frequency bands. Although the maximum transmission power allowed in different region could be different, in this evaluation,

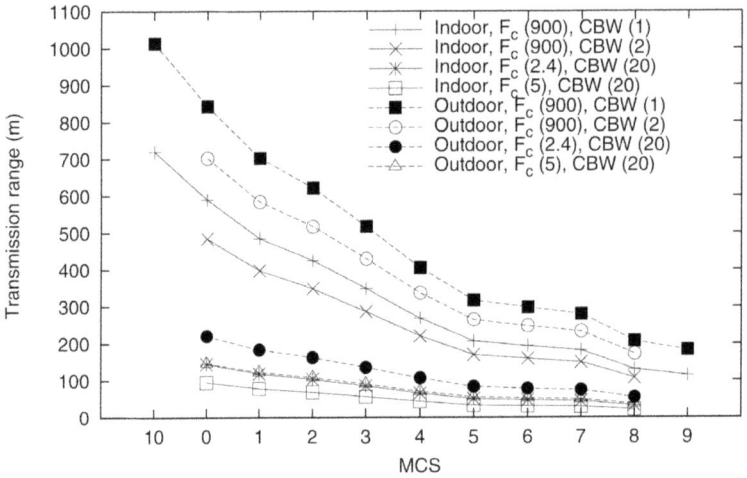

Figure 10 Comparison of transmission ranges with transmission power of 200 mW.

we set the transmission power to 200 mW, which is a typical transmission power in 5 GHz 802.11 system. The transmission range is calculated as the distance between transmitter and the receiver, of which the received power is equal to the minimum input level sensitivity specified in each system. The x-axis represents different MCS indexes, and the y-axis represents the transmission range. We compare the transmission range of 802.11ah, 802.11n's 2.4 GHz system, and 802.11ac's 5 GHz system, which are indicated here as $F_c(900)$, $F_c(2.4)$, and $F_c(5)$, respectively. Besides, regarding the channel bandwidth, we select 20 MHz channel bandwidth in 2.4 GHz and 5 GHz bands, and 2 MHz channel in 900 MHz band, which are indicated as $CBW(20)$ and $CBW(2)$, respectively. We additionally include the 802.11ah 1 MHz channel, which is denoted as $CBW(1)$. From the results, we observe that regarding the transmission range, the systems operating in outdoor environments generally outperform the systems in indoor environments as we can expect. We also conclude that the transmission range increases as the frequency of the operating band decreases due to the improved propagation characteristic of the wireless signal. One thing to note here is that MCS 10 and MCS 9 are only valid for 802.11ah's 1 MHz transmission in our results, since these MCSs are excluded by other cases according to the MCS exclusion rule of each system. Besides, the longest transmission range up to 1,013 m is achieved by the 802.11ah system with 1 MHz channel bandwidth in outdoor environment, which is almost

7 times of the 5 GHz system's outdoor transmission range shown in this result.

As the maximum transmission power level is different for different countries, we evaluate the 802.11ah transmission range with various transmission power levels. Figure 11 shows the increase of the transmission range as the transmission power increases from 200 mW to 1000 mW. We present the results obtained by transmissions in 1 MHz channel and 2 MHz channel in indoor and outdoor environments. In order to present the longest transmission range of each transmission power, we adopt the most robust MCS in 2 MHz channel, i.e., MCS 0. Besides, apart from MCS 0, we add MCS 10 in 1 MHz channel, which is only applicable to 1 MHz channel. In these results, when we set the transmission power to 1000 mW, the transmission range obtained by 1 MHz transmission with MCS 10 is 1,555 m, which is much longer than the required transmission range, i.e., 1,000 m. Moreover, the increase in the transmission power results in more significant increase in transmission range in outdoor environment than that in indoor environment.

### 6.2 Throughput Performance

We evaluate the throughput gain achieved by newly defined 802.11ah's medium access schemes described in this paper, i.e., RAW based channel access and speed frame exchange, through analysis. We also analyze the advantage of adopting the 802.11ah's compact frame formats.

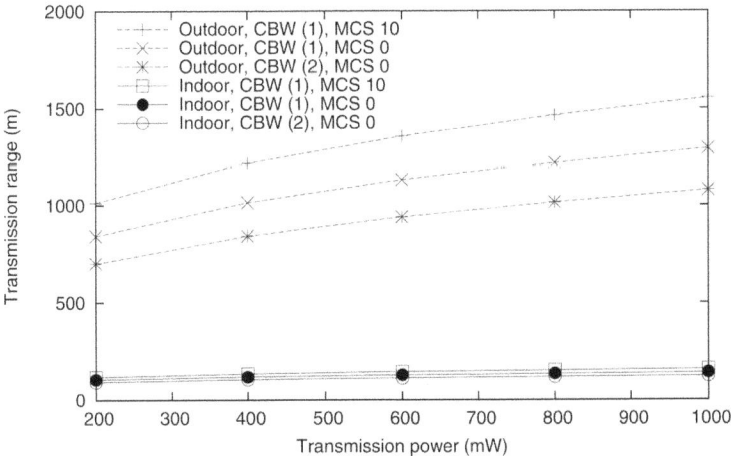

Figure 11 Transmission range of 802.11ah with different transmission power levels.

Figure 12 shows the analytical results. The x-axis represents MCS indexes and the y-axis represents MAC layer throughput for the case of a single transmitter and a single receiver without channel errors. We assume the transmission is conducted at 2 MHz channel with a single spatial stream as shown in Table 1. The legacy 802.11 channel access scheme, i.e., *Distributed Coordination Function (DCF)*, the RAW based channel access and the speed frame exchange are indicated as *Legacy*, *RAW* and *SFE*, respectively. When using 802.11ah compact frame formats, i.e., short MAC header and NDP ACK frame, the legend of each scheme is appended by *with compact format*, while it is denoted as *with normal format* otherwise. Besides, due to the traffic pattern of 802.11ah usage model, e.g., smart grid, we set the packet size to a relatively small value, which is 100 bytes in our evaluation. We assume that the MCS of the normal ACK frame is fixed to the most robust MCS, i.e., MCS 0, and the short GI is used for the evaluation.

The results show that speed frame exchange outperforms the other two schemes due to the eliminations of the channel access delay and ACK transmission. Similarly, the RAW based channel access, which is a contention-free channel access scheme, outperforms the legacy 802.11 MAC. Moreover, when using the compact frame format, the throughput achieved by each scheme increases as we can easily imagine. The throughputs presented in this evaluation results are much less than the data rates indicated in Table 1, since we set the packet size to a relatively small value so that the portion of the

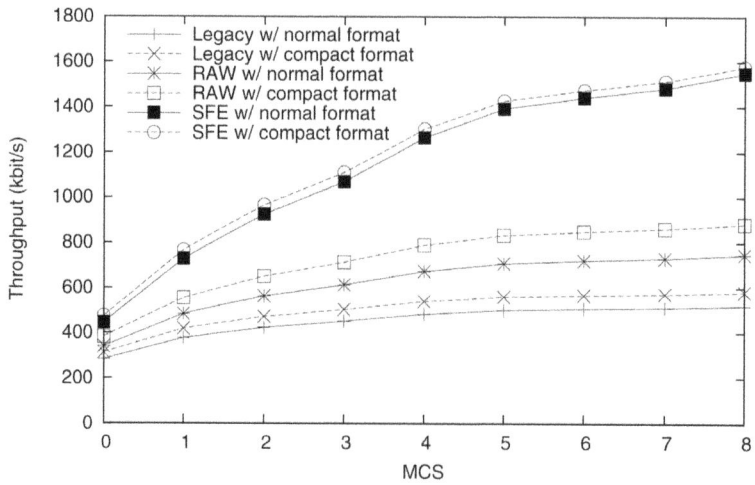

Figure 12 Throughput comparison of various MAC schemes.

protocol overhead, i.e., PHY header, MAC header, and ACK transmission, is relatively large. In our evaluation, when using speed frame exchange, the achieved throughput of MCS 8 is nearly 3 times of the throughput achieve by the 802.11 DCF with the same MCS.

## 7 Conclusion

In this paper, we introduce the 802.11ah project in terms of the use cases, PHY design, and MAC enhancements, for which the agreements are already made in drafting works of TGah. We also provide performance evaluation results in terms of transmission range and theoretical MAC layer throughput through theoretical analysis. The transmission range of 802.11ah is much longer than existing Wi-Fi systems, and the throughput can be significantly improved by newly adopted medium access schemes. The standardization of 802.11ah is still on-going, and to complete the standardization, some remaining issues need to be resolved, such as the performance degradation due to the interference caused by neighboring APs. In the near future, we envision that 802.11ah will be used for many emerging applications over very large scale wireless networks.

## References

[1] E. Perahia. IEEE 802.11n development: history, process, and technology. *IEEE Communications Magazine*, 46(7):48–55, Jul. 2008.
[2] IEEE std. IEEE 802.11ac/D5.0. Part 11: wireless LAN medium access control (MAC) and physical layer (PHY) specifications: enhancements for very high throughput for operation in bands below 6 GHz, Jan. 2013.
[3] D. Halasz. Sub 1 GHz license-exempt PAR and 5C. IEEE 802.11-10/0001r13, Jul. 2010, https://mentor.ieee.org/802.11/dcn/10/11-10-001-13.
[4] D. Halasz, R. Vegt. IEEE 802.11ah proposed selection procedure. IEEE 802.11-11/0239r2, Feb. 2011, https://mentor.ieee.org/802.11/dcn/11/11-11-0239-02-00ah-proposed-selection-procedure.docx.
[5] R. Vegt. Potential compromise for 802.11ah use case document. IEEE 802.11-11/0457r0, Mar. 2011, https://mentor.ieee.org/802.11/dcn/11/11-11-0457-00-00ah-potential-compromise-of-802-11ah-use-case-document.pptx.
[6] R. Porat, *et al.*, TGah channel model – proposed text. IEEE 802.11-11/0968r3, Jul. 2011, https://mentor.ieee.org/802.11/dcn/11/11-11-0968-03-00ah-channel-model-text.docx.
[7] M. Cheong. TGah functional requirements and evaluation methodology. IEEE 802.11-11/0905r5, Jan. 2012, https://mentor.ieee.org/802.11/dcn/11/11-11-0905-05-00ah-tgah-functional-requirements-and-evaluation-methodology.doc
[8] M. Park. Specification framework for TGah. IEEE 802.11-11/1137r14, Mar. 2013.

[9] E. Wong, et al., Two-hop relaying. IEEE 802.11-12/1330r0, Nov. 2012, https://mentor.ieee.org/802.11/dcn/12/11-12-1330-00-00ah-two-hop-relaying.pptx.
[10] G. Calcev, et al., Sectorization for hidden node mitigation. IEEE 802.11-12/0852r0, July 2012, https://mentor.ieee.org/802.11/dcn/12/11-12-0852-00-00ah-sectorization-for-hidden-node-mitigation.pptx
[11] NIST priority action plan 2. guidelines for assessing wireless standards for smart grid applications, ver. 1.0, Dec. 2010.
[12] IEEE Std. IEEE 802.15.4g-2012. Part 15.4: low-rate wireless personal area networks (LR-WPANs) amendment 3: physical (PHY) specifications for low-data-rate, wireless, smart metering utility networks, Apr. 2012.
[13] S. Aust, R.V. Prasad, and I.G. Niemegeers. IEEE 802.11ah: advantages in standards and further challenges for sub 1 GHz Wi-Fi. In *Proceedings of IEEE International Conference on Communications (ICC)*, Jun. 2012.
[14] Association of radio industries and business (ARIB), 950 MHz-band telemeter, telecontrol and data transmission radio equipment for specified low power radio station, english translation, ARIB STD-T96 Ver. 1.0, Jun. 2008.
[15] IEEE std. IEEE 802.11-2012. Part 11: wireless LAN medium access control (MAC) and physical layer (PHY) specifications, Mar. 2012.
[16] Further advancements for E-UTRA physical layer aspects, Annex A.2- system simulation scenario. Technical Report 36.814, 3GPP, Mar. 2010.
[17] V. Erceg, et al., TGn channel models. IEEE 802.11-03/940r4, May 2004.

## Biography

**Weiping Sun** received the B.E. degree of Network Engineering from Dalian University of Technology, Dalian, China in 2010. He is currently working toward a Ph.D. degree in the Department of Electrical and Computer Engineering, Seoul National University, Seoul, Korea. His current research interests focus on IEEE 802.11 WLAN MAC protocol and algorithm design.

**Munhwan Choi** received the B.S. and M.S. degrees in Electrical Engineering and Computer Science from Seoul National University, Seoul, Korea in 2005 and 2007, respectively. He is currently working toward a Ph.D. degree in the Department of Electrical and Computer Engineering, Seoul National University, Seoul, Korea. His current research interests include algorithmic design and protocol development for various communication systems such as IEEE 802.11 wireless local area networks and 60 GHz wireless personal area networks.

**Sunghyun Choi** is a professor at the Department of Electrical and Computer Engineering, Seoul National University (SNU), Korea. Before joining SNU in 2002, he was with Philips Research USA. He was also a visiting associate professor at Stanford University, USA from June 2009 to June 2010. He received his B.S. (summa cum laude) and M.S. degrees from Korea Advanced Institute of Science and Technology in 1992 and 1994, respectively, and received Ph.D. from The University of Michigan, Ann Arbor in 1999. His current research interests are in the area of wireless/mobile networks. He authored/coauthored over 150 technical papers and book chapters in the areas of wireless/mobile networks and communications. He has co-authored (with B. G. Lee) a book entitled "Broadband Wireless Access and Local Networks: Mobile WiMAX and WiFi," Artech House, 2008. He holds about 100 patents, and has tens of patents pending. He is also currently serving on the editorial boards of IEEE Transactions on Mobile Computing and IEEE Wireless Communications. He has received a number of awards including the Presidential Young Scientist Award (2008); IEEK/IEEE Joint Award for Young IT Engineer (2007); Shinyang Scholarship Award (2011); the Outstanding Research Award (2008) and the Best Teaching Award (2006) from the College of Engineering, SNU; and the Best Paper Award from IEEE WoWMoM 2008.

# Towards Standardized Prevention of Unsolicited Communications and Phishing Attacks

JaeSeung Song and Andreas Kunz

*NEC Laboratories Europe*

Received 3 April 2013; Accepted 14 May 2013

## Abstract

The world of communication technology is changing fast and the means of communication are moving towards a packet switched transmission systems such as Voice over IP (VoIP). Formerly call identity spoofing of the displayed number in circuit switched (CS) networks was too difficult to perform so that people could be sure that when receiving a call on their mobile phone or at home, the displayed number is the one as it is supposed to be. Nowadays this is not the case anymore, voice communication from the internet with VoIP is cheap and spam calls can be easily realized without any costs, also it is getting easier to perform spoofed calls with wrong display name or number.

The mobile network operators have no mechanisms to tackle those threats, but standardization activities are already in place within the security group SA3 of 3GPP. This paper provides an overview of the current status of the standards activities and shows the most promising solutions that are proposed up to now. The proposed solutions detect unsolicited communications and spoofed calls by tracing back to the displayed number used in the attack.

**Keywords:** Spam, Unsolicited Communication, Voice Phishing, Call Id Spoofing.

## 1 Introduction

Unsolicited communication (UC) is defined bulk voice communication in communication networks where the benefit is weighted in favor of the sender [1]. Due to its anonymous, low cost and easy-to-use applications, unsolicited communication has become a popular method used by attackers [23]. Although many solutions protecting unsolicited communication exist [11], the volume of spam emails and the amount of financial damages are increasing rapidly every year [24]. Protecting users from unsolicited communication is now an important topic among network operators and system vendors, because it enables to provide high-quality services to users and helps to reduce management costs. However, this requires collaborations between stakeholders such as customers, operators, system vendors and legitimate organizations. To tackle this problem, various standards organizations (SDOs) have started their studies in this area.

The remainder of this article is organized as follows. The next section gives background information about security threats in unsolicited communications. Section 3 provides various SDOs' latest standardization activities related to unsolicited communications together with a potential solution. In Section 4, we focus our attention specifically on voice spoofing attacks and describe latest standardization activities. We also present our prototype implementation. After that the paper finishes with conclusions in Section 5.

## 2 Security Threats in Unsolicited Communications

The introduction of low-cost communications to operators' networks, such as VoIP and IP Multimedia Sub-system (IMS) [17], imposes many security threats by unsolicited communications, as listed below. Compared to the traditional voice networks, e.g., Public Switched Telephone Networks (PSTN), network operators provide significantly low-cost VoIP based services such as SMS, email and voice call. The customers have been enjoying such services because they are very attractive from the cost aspect and provide various rich multimedia services. On the other hand, this makes VoIP an attractive carrier for delivering unsolicited communications by spammers or attackers.

There exist security mechanisms provided by such networks or services, however, customers are still suffering from several types of security threats, for instance, fraud, spam emails and voice phishing attacks. We describe several well-known security threats resulting from unsolicited communication as follows:

- Flooding attack [20]: The attacker can generate a large number of unsolicited messages and send them to victims such as terminals and network nodes.
- Spam-over-Internet-Telephony (SPIT) [21]: This attack is similar to E-mail spam. The attacker disturbs users or group of users through placing unsolicited calls. Since such unsolicited calls can be made using bots or other malicious software at any time (for instance at the midnight) with extremely low costs, many network operators are seeking a solution for protecting unsolicited communication in their VoIP networks.
- Private information leakage [22]: A victim may receive a spoofed call with a call ID that is known by the victim to be the one from his bank and the victim may provide private information or passwords about his bank account. Conversations and messages can easily be intercepted by attackers, e.g. in case he mobile operator uses the call ID for providing access to the voice mailbox, which lead the leakage of private information.

Although existing solutions provide a certain level of protections to customers, they often fail to detect such threats because legislation issues and frequent changes of threat patterns [25].

The threats listed above usually harm user experience and cause annoyance for users. On the other hand, there exist other types of threats used for taking a monetary benefit from the user, which is called "voice phishing". Typically, attackers modify the caller ID, i.e., displayed telephone number, of incoming call and pretend themselves as a trustworthy person or legitimate organizations such as bank [20]. This caller ID spoofing can easily be made through various methods, for instance using a VoIP client or spoofing web sites. Since the damage caused by phishing attacks has been increasing steeply in the recent years [18], there is a strong need for protecting customers against voice phishing attacks.

## 3 Prevention of Unsolicited Communications

The growth of the bandwidth capacities in the networks due to the demand of more and more resource hungry applications and the highly competitive market situation of the network operators led to a significant decrease of the prices for data connectivity services. Now it became very cheap and easy for attackers to distribute unsolicited voice communication, since only a SIP-Server is needed with nearly no costs in distributing the messages to a huge amount of (randomly) selected recipients. There are several services that can be realized with SIP such as multimedia video, voice, messaging, etc. standardized by the Internet Engineering Task Force (IETF). 3GPP then further reused the IETF work and created services for operators in the IP Multimedia Subsystem (IMS) [17], which can be used as well by fixed and mobile network operators. All those services can be misused for unsolicited communication.

## 3.1 Status of Standardization

Since unsolicited communication is becoming a more and more common issue for operators around the world [25], several standardization organizations already looked into the problem and tried to address it for selective services. [18] gives a short overview on the past standardization activities. IETF discussed several internet drafts about SPam over Internet Telephony (SPIT), but all of them expired and no RFC was created. There is no active work in IETF on this topic. ETSI TISPAN performed two studies [5, 6] on prevention of unsolicited communication in the Next Generation Network (NGN). The security group of ITU-T produced several recommendations ([12–16]), analyzing the different types of unsolicited multimedia communication. The recommendations on the overall aspects [15] proposes corresponding countermeasures, but only limited to authentication, authorization and security management. The technical strategies [12] and the overall framework [16] differentiate between store-and-forward and real-time communication for the type of the service. The technical strategies propose a hierarchical model with filtering strategies, feedback strategies [12], service strategies, equipment strategies and network strategies. The framework consists of anti-spam functions on sender, core and recipient side, which can perform several actions e.g., protocol analysis and filtering. GSMA focus with their recommendation on call ID spoofing and phishing attacks, which are the main frauds in the mobile networks, but with the introduction of IMS for voice and multimedia services also other unsolicited multimedia communication will increase. 3GPP is the only major standardization organization that is at the moment still trying to find a solution for the prevention of unsolicited communication and call spoofing attacks. Two studies were carried out for Prevention of Unsolicited Communication in IMS (PUCI) ([1, 2]) and a new study is actively discussed on the prevention of caller ID spoofing [19]. Even the two studies on PUCI did not lead to normative work up to now; nevertheless the findings of the work are worthwhile to be described further in more details.

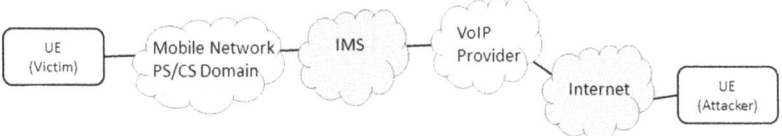

Figure 1 UC Source outside the operator network

## 3.2 Scenarios

3GPP analyzed in [1] the two basic scenarios of the threat of unsolicited communication (UC), i.e., the source of the UC is:

- Inside the mobile operator home network of the victim
- Outside the mobile operator home network, as shown in Figure 1

Depending on the location of the UC source, different network nodes are impacted in order to host the functionality to prevent that the UC is successfully established to the victim and to block the UC setup attempts. There are many accompanying threats e.g., like cost creation if the victim has supplementary services enabled like call forwarding, victim is roaming, phishing, equipment hijacking etc.

All communication services available in IMS were considered as potential source of UC, therefore all solutions analyze the SIP signaling and treat the session accordingly.

## 3.3 Available Solutions and Analysis

There are two main solutions that can be applied effectively to fight UC in the network: one based on supplementary services and one based on identification, marking and reacting to the UC session (IMR).

### 3.3.1 UC Protection with Supplementary Service

Using supplementary services is from deployment perspective the easiest way to provide a limited protection against UC to the end customers. Supplementary services are usually hosted in the Telephony Application Server (TAS). Figure 2 shows a simple architecture where the attacker and the victim are located in the same IMS network:

The user devices, called User Equipment (UE) connect to the proxy server (Proxy Call Session Control Function, P-CSCF) in the IMS network. The P-CSCF itself connects to the Serving Call Session Control Function (S-CSCF), which provides originating and terminating services with the help of

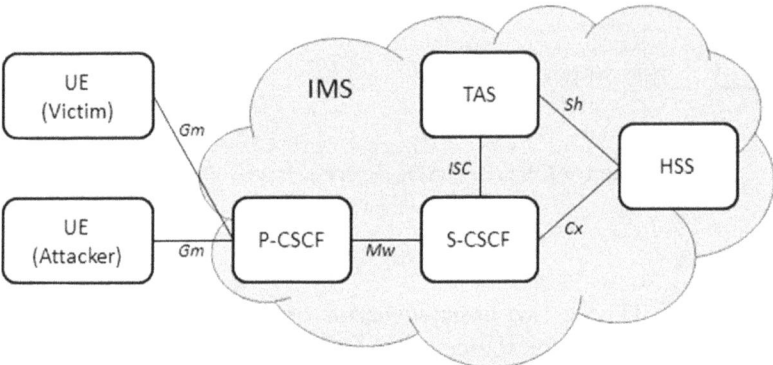

Figure 2 : Simple IMS Network Architecture

Application Servers (AS) based on the subscriber profile, downloaded from the Home Subscriber Server (HSS) at the time of registration. The IMS network itself is access agnostic and can provide multimedia services for fixed and mobile networks.

There are several supplementary services that are suitable for filtering incoming session requests:

- Incoming Call Barring with White List: the caller's ID is compared with the white list and the session setup is interrupted if the calling ID is not on the list
- Incoming Call Barring with Black List: the caller's ID is compared with the black list and the session setup is interrupted if the calling ID is on the list
- Anonymous Call Rejection: session setups without calling ID, i.e., restricted asserted public user ID, are rejected
- Closed User Groups: special trust network based on white list
- Call Diversion on Originating Identity: the callee can redirect the incoming session, e.g., to a mailbox
- Malicious Customer Identification: generates a trace of the last anonymous session to identify the source

All these supplementary services can be of course also combined, the disadvantage of such a solution is that the configuration is not updated in real time, e.g., customers configure their black and white lists based on their experience and normal call behaviour. If e.g., an attacker creates outside the network random public user ids for unsolicited communication, the black listing filtering would be not successful. Using white lists would prevent this, but this has the drawback that the user can be called only by the very limited and comparable small group of people on the list. For this reason another solution is described in the next chapter, which overcomes those problems.

### 3.3.2 UC Protection with IMR

In order to be able to dynamically react on unsolicited communication attacks, those sessions need to be identified, marked for further processing and afterwards a decision is taken to react to the attack. This concept is called IMR (Identification, Marking, Reaction).

The incoming session request is then processed according to the three stages:

- Identification: the UC identification can be classified into three categories:
  
  o Non-intrusive tests: analysis of session signaling
  o Intrusive tests: caller test to identify UC attempt
  o Feedback by user: personal black list, react during call etc.

- Marking: The session gets marked with a UC score to rate the session.
- Reaction: based on the UC score different actions can be performed, e.g., blocking the session, redirection to mailbox, automatic update of filter lists etc.

Figure 3 shows the simplified IMS architecture with attacker and victim in the same network and with IMR functionality in the Application Server (AS) and in the S-CSCF. The AS could also interact with a content inspection function in order to test the incoming session, e.g., by playing an announcement at the Media Resource Function (MRF) to press certain keys and then to analyze the DTMF answer from the caller. If the caller would be an attacker who does random calls in order to play e.g., commercials then it cannot answer the test and would be classified as UC.

The S-CSCF could also do some simple testing, e.g., analyzing the session setup rate from a specific source and then mark the sessions accordingly.

All tested sessions are marked with the UC score, the result of the tests. There may be only one test, but there could be also different tests in sequence and the UC score would be updated accordingly. If the UC score is transmitted between operators, then it would be beneficial also to agree for the UC score in the Service Level Agreements (SLA) on the range and the threshold, i.e., the UC score from when onwards an operator considers a session to be UC.

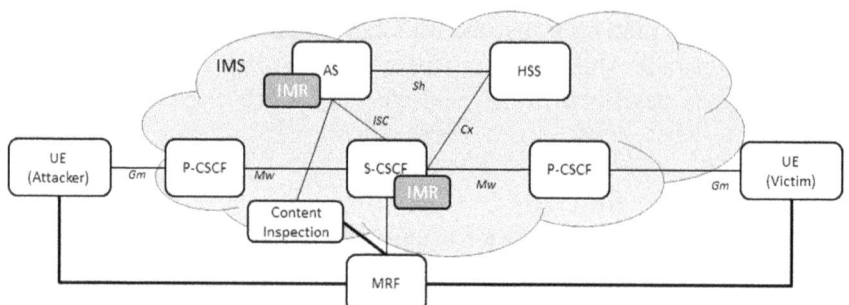

Figure 3 : Simplified IMR Architecture

Incoming session requests that got already tested and marked in the originating network can then be mapped to the UC score used in the intermediate or terminating network. The terminating operator can then decide how to react on the session request.

### 3.4 Proposed Solution

Operators should start and take countermeasures to the increasing problem of UC attacks in the networks. Starting with a basic solution with Supplementary Services that provides some elemental protection of the subscribers, it is recommended to use an IMR system to test incoming and outgoing session requests. Only with an IMR solution it is possible to identify also more sophisticated UC session requests and to react dynamically to them. Additionally the IMR system is learning and can dynamically update filter lists according to the tests.

## 4 Prevention of Voice Phishing Attacks

Voice phishing called vishing is a scam usually carried out by unsolicited communication in particular using voice call to obtain sensitive information from users, such as login credentials or information to be used for identity theft. The main objective of the attacker is usually to gain monetary benefits from victims. In this section we provide an overview of latest standard activities of the two major SDOs (i.e., 3GPP and TISPAN) together with potential solutions to prevent voice phishing attacks. We also introduce a prototype implementation to show the feasibility of the proposed solution in Section 4.3.

## 4.1 Use Cases and Attack Scenarios

So far most work that has been done in 3GPP regarding spoofed call detection and prevention is about analyzing various use cases. PUCI TR [1] addresses two popular use cases which are often used by attackers for phishing. There exist various scenarios achieving these use cases. In this section we introduce these use cases together with some popular phishing attack scenarios.

**Use cases:** A main purpose of phishing attacks is for financial gain of the attacker. The attacker usually tries to get sensitive information from users such as bank account information and login credentials.

The first use case is about the *leakage of personal bank account information*. Similar to email phishing scams, the attacker approaches users using phone calls pretending to be from the bank or the government organization. The attacker then asks victims to disclose their bank account information or transfer money. Sometimes the attacker even tries a prior call where no information is required in order to convince victims that the call is from a legitimate bank. The user is then easily fooled by receiving a subsequent call, which refers to the initial call.

Secondly, identity theft is introduced as another popular voice phishing use case. The attacker aims to get personal information from a victim, and the information is then used to obtain credit in the name of the victim. One popular example, for instance, is to call a user and saying the user has won a prize. The user is asked to provide certain sensitive personal information to collect the prize.

**Attack scenarios:** The attacker uses various methods to trick users to make them believe a call is from a legitimate company or organization. Figure 4 shows four common voice spoofing scenarios, and they are described in the following:

1. **IMS Application Server**: Within IMS, application servers acting as a back-to-back user agent (B2BUA) can be deployed by $3^{rd}$ party service provider. Such application servers can easily change the identities of the incoming SIP request and initiates a new one with faked ID towards the victim.
2. **Private Branch eXchange (PBX)**: In a typical telecommunications system, a Primary Rate Interface (PRI) is used to establish a connection between a PBX and a local network. Such PRI trunks are generally trusted by the network operator, and any caller ID through these trunks are delivered to the user without verifications.

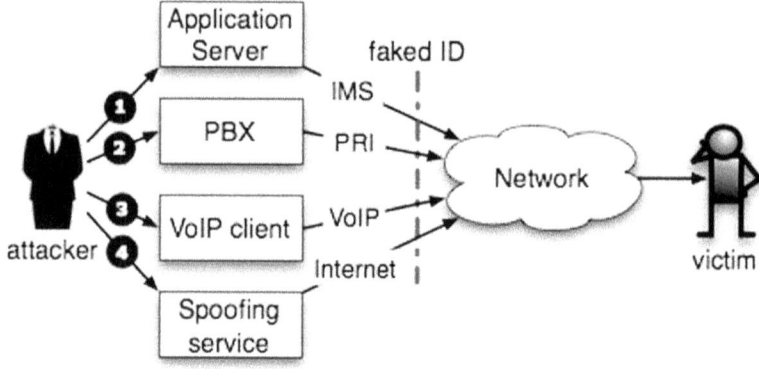

Figure 4 Common voice spoofing scenarios

3. **VoIP client**: Using VoIP clients is the easiest way to generate spoofed calls. There exist many VoIP clients that allow the attacker to attach a spoofed caller ID to the destination field of the data packet. For instance, in the SIP protocol [8], caller ID is provided by the "From" header of a SIP message in requests.
4. **Caller ID Spoofing service**: There are many online web sites providing a caller ID spoofing service. The attacker can easily subscribe such service and modify the caller ID. In this case, the faked caller ID is displayed on the victim's UE to a legitimate entity such as bank and policy station.

### 4.2 Available Solutions and Analysis

There exist many different proposed solutions to protect users from voice phishing attacks. They can be roughly categorized into three types: (1) *voice analysis*, (2) *blacklist & whitelist* and (3) *runtime ID checks*. Each technology is described in the followings.

**Voice analysis:** Several solutions recently proposed introduce a mechanism analyzing an incoming voice call to find a pattern that can distinguish spoofed calls from normal calls. PinDr0p [7] assists users to guess the source and the path taken by a call through analyzing network specific characteristics such as packet loss, noise profiles and applied voice codecs. It is possible to use an algorithm based on Gaussian mixture model. Chang et al. [9] use the fact that the human voice can be used for detection of deception. For instance, the voice of a liar usually has a larger pitch lag value than the normal voice.

Chang et al. first extract coding parameters followed by selecting relevant feature vectors to detect voice phishing.

Although such algorithms can easily be implemented in the user's terminal, the main problem is that analyzing codecs and characteristics of the call can require significant processing power. In addition, these voice analysis methods would be unable to cope with all different kinds of phishing attacks.

**Blacklist & whitelist:** The use of a blacklist (or a whitelist) [10] which can detect previously known phishing (or legitimate) caller IDs can reduce the traffic usage by filtering phishing attacks at the earliest possible stage; that is, before forwarding them to the callee. Surely, it would be difficult to maintain the latest blacklists (or whitelists) either on mobile phones or on the database deployed in network entities. There also exist several legal issues that need to be considered by network operators when rejecting incoming calls from certain user accounts. For example, there will always be countries where it is legal to send SPIT.

**Runtime ID checks:** When spoofed calls are delivered to end users, they usually do not have enough information to judge that the caller ID is spoofed. On the other hand, the first entry point to the operator network has a lot more information. This first entity can be used to initiate a verification process of the originating party caller ID to check whether there is an ongoing call to the request caller ID. Although this requires an enhancement to an interworking gateway (i.e., the entry point of the operator network), such method provide several advantages over others, including no impact on call setup time and performance.

## 4.3 Proposed System Implementation

Since VoIP is a real-time communication, we believe that a method checking the caller ID at runtime is a promising solution to detect phishing attacks while avoiding many drawbacks. Since different players, such as the mobile network operator, entities that want to be trusted (banks, governments, etc.) and customers, are involved in providing the protection of voice phishing attacks, standards are required to define information exchange procedures. This section describes a system that we developed and implemented to provide runtime caller ID verification.

We introduce a system that detects possible voice phishing attacks through checking the display name of an incoming call at runtime. First, the system uses the fact that the display number is faked in a spoofed call and subject for the verification. Second, the system performs the verification process at runtime either initiated by user on-demand or as a supplementary service. A method used in the system traces back through the incoming call routes to the entity that operates the actual caller ID in order to detect spoofed calls.

Figure 5 shows a simplified architecture together with the caller ID verification procedure. To know a caller is using a faked caller ID, the callee initiates the caller ID verification process by asking the interworking GW (Ingress Entity) to check (1, 2). When the GW is asked by the user (3), it formulates a verification request message with the display number and forwards the message to the actual organization that owns the display number (4). The organization then checks its registered subscribers whether any of them are using the faked display number (5). The verification results are then reported to help the user decides the call is spoofed (6).

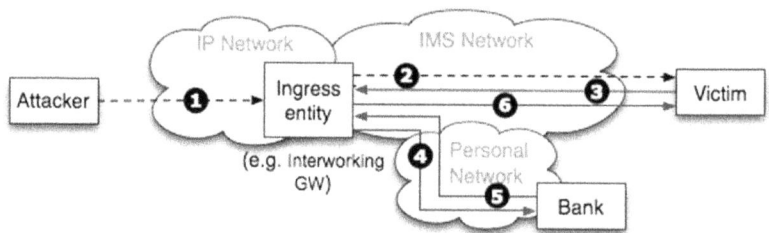

Figure 5 A potential high level architecture

## 5 Conclusions

This paper has provided an overview of standardization activities associated with preventing unsolicited communications. Unsolicited communications, such as spam emails and voice phishing attacks, are becoming a serious problem for both users and network systems. Therefore, studies and specifications in various SDOs have gained broad industry attention and support. Most SDOs, such as 3GPP, ITU-T, TISPAN, etc., have completed their study on the analysis of unsolicited communications and are now considering to start normative work to standardize a solution for protecting unsolicited communication attacks.

After introducing several existing solutions, we proposed potential frameworks in Sections 3.4 and 4.4 to mitigate the threats from both unsolicited communications and call spoofing attacks, respectively. We show that these solutions easily can be introduced to the existing network architectures while having minimal impact to the current network architecture and network design.

As a future work, we intend to integrate two proposed systems into a generic UC protection system in order to reduce complexities and maintenance cost. For instance, through combining the proposed systems in Sections 3.4 and 4.4, we can manage a single unified blacklist/whitelist for all UC calls.

## References

[1] 3GPP TR 33.937 "Study of mechanisms for Protection against Unsolicited Communication for IMS (PUCI)"
[2] 3GPP TR 33.838 "Study on Protection against Unsolicited Communication for IMS (PUCI)"
[3] 3GPP S3-121245 "Security study on spoofed call detection and prevention;(Release 12)"
[4] 3GPP TS 24.416 "TISPAN; PSTN/ISDN simulation services; Malicious Communication Identification (MCID); Protocol specification"
[5] ETSI TR 187 015 Ver. 3.1.1, "Telecommunications and Internet converged Services and Protocols for Advanced Networking (TISPAN); Prevention of Unsolicited Communication in the NGN"
[6] TR 187 009 Ver. 2.1.1 "Telecommunications and Internet Converged Services and Protocols for Advanced Networking (TISPAN); Feasibility study of prevention of unsolicited communication in the NGN"
[7] Balasubramaniyan, V.A., Poonawalla, A., Ahamad, M., Hunter, M.T., Traynor, P.: Pindr0p: using single-ended audio features to determine call provenance. In: Proceedings of the 17th ACM conference on Computer and communications security, CCS '10, pp. 109-120. ACM, New York, NY, USA (2010).
[8] Rosenberg, J., Schulzrinne, H., Camarillo, G., Johnston, A., Peterson, J., Sparks, R., Handley, M., Schooler, E.: SIP: Session Initiation Protocol. RFC 3261 (Proposed Standard) (2002). URL http://www.ietf.org/rfc/rfc3261.txt
[9] Chang, J.H., Lee, K.H.: Voice phishing detection technique based on minimum classification error method incorporating codec parameters. Signal Processing, IET 4(5), 502-509 (Oct.)
[10] Kolan, P., Dantu, R.: Socio-technical defense against voice spamming. ACM Tranactions on Autonomous and Adaptive Systems (TAAS). 2(1) (2007)
[11] Schmidt, A., Leicher, A., Shah, Y., Cha, I., Guccione, L.: Sender scorecards for the pre vention of unsolicited communication. In: Collaborative Security Technologies (CoSec), 2010 IEEE 2nd Workshop on, pp. 1-6 (2010)
[12] X.1231 Technical strategies for countering spam
[13] X.1242 Short message service (SMS) spam filtering system based on user-specified rules
[14] X.1243 Interactive gateway system for countering spam
[15] X.1244 Overall aspects of countering spam in IP-based multimedia applications
[16] X.1245 Framework for countering spam in IP-based multimedia applications
[17] 3GPP TS 23.228 "IP Multimedia Subsystem (IMS); Stage 2"
[18] Nico d'Heureuse, Jan Seedorf, Saverio Niccolini, Thilo Ewald: Protecting SIP-based Networks and Services from Unwanted Communications. IEEE "GLOBECOM" 2008

[19] 3GPP TR 33.8de "Security study on spoofed call detection and prevention; (Release 12)", S3-130242
[20] Keromytis, A.: A survey of voice over ip security research. In: A. Prakash, I. Sen Gupta(eds.) Information Systems Security, Lecture Notes in Computer Science, vol. 5905, pp. 1 – 17. Springer Berlin Heidelberg (2009)
[21] Quittek, J., Niccolini, S., Tartarelli, S., Schlegel, R.: On spam over internet telephony (SPIT) prevention. Communications Magazine, IEEE 46(8), 80 – 86 (2008)
[22] Neumann, T., Tillwick, H., Olivier, M.: Information leakage in ubiquitous voice-over-ip communications. In Trust and Privacy in Digital Business, Lecture Notes in Computer Science, vol. 4083, pp. 233-242. (2006)
[23] Nassar, M., Niccolini, S., State, R., Ewald, T.: Holistic voip intrusion detection and prevention system. In Proceedings of the $1^{st}$ international conference on Principles, systems and applications of IP telecommunications, IPTComm '07, pp. 1-9 (2007)
[24] J.Rosenberg and C. Jennings, "The Session Initiation Protocol (SIP) and Spam" IETF RFC 5039, jan. (2008)
[25] Frost, N.: VoIP: VoIP threats – getting louder. Netw. Secur. 2006 (3), 16-18 (2006)

## Biography

**JaeSeung Song** is currently working as a senior researcher and oneM2M standardization engineer at NEC Europe Ltd, Heidelberg, Germany. Previously, he worked for LG Electronics as a senior research engineer from 2002 to 2008. He received a PhD at Imperial College London in the Department of Computing, United Kingdom and BS and MS degrees from Sogang University.

**Andreas Kunz** received his diploma degree and his Ph.D. in Electrical Engineering from the University of Siegen, Germany. He is working for NEC Laboratories Europe with focus on 3GPP standardization, mainly in the system architecture working group.